Agronomists and Food: Contributions and Challenges

ASA Special Publication Number 30

Papers presented at the annual meeting of
the American Society of Agronomy in
Houston, Texas, November 28—December 3, 1976

Editor: **Marlowe D. Thorne**
Editor-in-Chief: **Matthias Stelly**
Managing Editor: **David M. Kral**
Assistant Editor: **Judith H. Nauseef**

1977

Published by the

AMERICAN SOCIETY OF AGRONOMY
677 South Segoe Road
Madison, Wisconsin 53711

Library of Congress Catalog Card Number: 77-6065
Standard Book Number: 0-89118-048-6

Contents

Foreword

The annual meeting of our Societies in our nation's Bicentennial year provided an appropriate forum for reviewing the contributions of agronomists and for examining the challenges ahead. This Special Publication will provide a lasting record of many of the special addresses given at the 1976 annual meeting. This publication takes its title from the theme of the meetings. It focuses both the past and the future on mankind's greatest need—food for the world's ever-growing population.

Included in this Special Publication are the Keynote Address, the two papers given in the ASA Special Session and the three in the CSSA-SSSA Joint Session. In addition there are three of the Landmark or Bicentennial papers presented in some ASA and CSSA division commemorative sessions by outstanding members of the division. Thus we have in this volume both general and specific reviews of our professional history and achievements and an abundance of challenges for the years ahead.

The officers of the American Society of Agronomy are proud to present to our members, colleagues, and friends this Special Publication which we hope will serve as an inspiration and a stimulus to all who are truly concerned about adequate food for the people of the world.

Marlowe D. Thorne, President
American Society of Agronomy

Ecology and Agricultural Development: Striking a Human Balance

Clifton R. Wharton, Jr.

INTRODUCTION

The problems of providing food for the starving and undernourished millions in the world are still with us. Despite the intermittent interest of the news media in the subject, incipient famine and pervasive malnutrition continue to require our compassionate professional attention.

Statistics are a rather cold way to describe this very human problem, but they clarify the issue. It is estimated that one out of six persons in the world is undernourished. Most of these people live in the developing countries where food production is still insufficient. Population size, the major determinant of the demand for food, is growing faster in these countries, which already account for 70% of the world's present population, than in industrialized countries.

As a result, this is a period of great anxiety about the world's ability to feed its growing population. In the early 1970's, we went from a state of food surpluses and low prices to one of relative food scarcity and high prices. This rapid reversal has raised again a wave of widespread pessimism about the ability of food production to keep pace with population growth.

To be sure, progress has been and is being made. Countless lives have been saved through technical assistance, new technologies, and economic aid. The agronomists, crop scientists, and soil scientists who have contributed to so many past successes in achieving greater food production should feel justifiable pride. Whether they worked on high lysine corn (*Zea mays* L.), short-stem wheat (*Triticum aestivum* L.), or high-yielding rice (*Oryza sativa*), these scientists have helped to hold at bay the Malthusian specter. But, despite all that has been accomplished in the past, the problem of explosive population growth is no less acute.

When looking to the future and the role of agricultural professionals, it is clear that the challenges remain and may well become greater. There is something grimly amusing in the constantly changing views of experts on the

---

Clifton R. Wharton, Jr. is president of Michigan State University in East Lansing, Mich.

dimensions of our prospective food crisis. As the apocryphal old professor was supposed to have said, "The questions don't change, but the answers do." For example, the first Club of Rome report of 1970 made us believe that doomsday was right around the corner. Its study, *The Limits to Growth* (Meadows et al., 1972), stated that the earth does not have sufficient resources to sustain the rate of world economic growth experienced over the past 70 years. Even if we could continue the pace, the environment could not absorb the accompanying pollution.

In 1976, the Club of Rome issued its second report, which stated that the impending decline in the human race might not be as bad, or as rapid, as its earlier prediction. In the same vein, that well-publicized cosmic thinker, Herman Kahn (Epstein, 1976), has recently assured us that all of these predictions of doom are weak and that the future is serious but not so bleak.

The debate over the projections seems to me to deflect attention from the critical point: no matter which estimates are accepted, we face massive food problems over the next 25 years. The world's population today is over 4 billion souls. Population experts tell us that this number will double in the next 25 years, and that by the year 2000, there will be 8 billion people.

It is difficult to conceptualize the enormity of these prospective developments. It took us from the beginning of time to the year 1650 for the world to achieve a population of 500 million people. In the next 200 years, the population doubled and achieved the number of one billion. That was roughly in 1850. Only 80 years later, in 1930, the number had doubled again and reached 2 billion. By last year, this total had doubled again to 4 billion people, and by the year 2000, a mere 25 years later, it will reach 8 billion.

The world is adding 70 million persons a year, over 200,000 a day. Some 60 million of the 70 million annual increase will occur in the less-developed areas. For instance, in 25 years, Asia's population alone will be more than the total world population today! The significance of this population explosion needs no explanation to scientists and researchers whose entire livelihood is devoted to the production of food—the basic nutrient for the sustenance of human life.

When these awesome statistics are placed in the wider context of the other factors which are shaping our daily lives, such as the tremendous pressures upon our limited natural resources, rising incomes translated into larger demands for the better things of life, and the escalating shortages in and costs of purchased farm inputs, there is little doubt that the issue of food will be a major challenge facing the world for the next 25 to 50 years.

TITLE XII, THE FINDLEY-HUMPHREY AMENDMENT

Fortunately, we in the United States have not been found wanting in our willingness to meet this challenge. One new approach has been launched which offers an unprecedented vehicle to attack the problem. I refer to

Title XII, Famine Prevention and Freedom from Hunger, or the Findley-Humphrey Amendment to the Foreign Assistance Act of 1975.

The central intent of Title XII is to promote an expanded role for U. S. agricultural colleges and universities in helping to solve the critical food problems of the developing world. The legislation is based upon the belief that much of the success of U. S. agriculture is due to the combined approach of teaching, research, and extension in our agricultural colleges and universities.

Supervising the Title XII effort is the Board of International Food and Agricultural Development, which is composed of seven persons, at least four of whom must be from land-grant institutions. I was honored to be appointed chairman. The other members of the Board are Gerald W. Thomas, President, New Mexico State University; Orville G. Bentley, Dean of Agriculture, University of Illinois; Anson R. Bertrand, Dean of Agriculture, Texas Tech University; Charles Krause, President, Krause Milling Company, Milwaukee, Wis.; and James J. O'Connor, private consultant, Houston, Tex. (The seventh position will be filled shortly.)

The Board has been given significant authority. It is charged with assisting the Agency for International Development (AID) in planning and implementing all international food and nutrition programs. The Board will also assist AID in making an annual report to the Congress on the past year's programs, and on activities projected five years into the future.

AID has long used the expertise of U. S. universities and colleges in foreign assistance programs. Some 143 universities currently hold AID contracts. But the new authorization permits more systematic and longer-term application of their scientific and technological expertise in developing countries. Universities will be encouraged to integrate their overseas and domestic efforts to achieve maximum feedback of the overseas experience into their own research, teaching, and public service responsibilities.

Assistance is to be provided in five broad areas as written into the amendment. They are

(1) to strengthen the capabilities of universities in teaching, research, and extension work to enable them to implement current programs authorized. . .and those proposed. . .;
(2) to build and strengthen the institutional capacity and human resource skills of agriculturally developing countries so that these countries may participate more fully in the international agricultural problem-solving effort and to introduce and adapt new solutions to local circumstances;
(3) to provide program support for long-term collaborative university research on food production, distribution, storage, marketing, and consumption;
(4) to involve universities more fully in the international network of agricultural science, including the international research centers, the activities of international organizations such as the United Nations Development Program and the Food and Agriculture Organization, and the institutions of agriculturally developing nations; and

(5) to provide program support for international agricultural research centers, to provide support for research projects identified for specific problem-solving needs, and to develop and strengthen national research systems in the developing countries.

(Section 297 (a) of Title XII)

The most significant feature of the new legislation is that it mandates direct participation by U. S. universities, through the Board, in the total process. The Board's duties include: (i) participating in the formulation of basic policy, procedures, and criteria for project selection, implementation, and evaluation; (ii) developing a roster of eligible universities; (iii) recommending which developing nations could benefit from programs; (iv) reviewing and evaluating the terms and conditions under which the universities participate in the program; (v) reviewing and evaluating their agreements; (vi) recommending to the AID Administrator how funds are to be apportioned for these purposes; and (vii) assessing the impact of the programs carried out under Title XII.

As was recently observed by Ralph Huitt, the Executive Director of the Land-Grant Association:

> . . .Never before has an organized state put such a burden for carrying out national policy on institutions of higher education as the United States does here, and never before has the credibility of a certain kind of institution and its mode of operations been required to stand so severe a test. The government has said in effect: "All right, you have proved you can teach Americans how to feed themselves. Now prove you can teach the people of the world how to feed themselves." . . . Opportunity knocks–and it's enough to scare hell out of anybody with any sense.[1]

Another central feature of the Act is its recognition that the capabilities of U. S. universities must be strengthened on a long-term basis if the purposes are to be achieved. There is little need to explain to agronomists, crop scientists, and soil scientists that successful agricultural research and extension, whether at home or abroad, require long-term support. Short-term, erratic funding will not do. Without a strong, sustained, long-term commitment of talent and energies, Title XII will not be successful. The resources and experience are available on our campuses, but they must be forged into a critical effective mass. This can only be done by assured long-term support.

Recently, we were treated to stories from Washington, D. C. that some $11 to $17 billion, most of it appropriated to the Department of Defense, did not get spent during the just-ended fiscal year. The bureaucrats were said to be "embarrassed" by their inability to handle these riches.

I know it is overly simplistic to suggest that funds earmarked for one purpose can readily become available for another. But the magnitude of what Washington calls a "shortfall" makes one speculate on what could be achieved if just $1 billion could be committed to Title XII for institutional

[1] Ralph W. Huitt, Report of the Executive Director to the Senate, National Association of State Universities and Land-Grant Colleges, 14–17 Nov. 1976.

strengthening over a 10-year period. If half the funds were committed to programs focused on the long-run strengthening of U. S. institutions and half for foreign institutions, for example, this could provide $1 million to each state each year over the 10-year span. A modest amount, indeed, when compared with the $14 million cost of the fighter plane that fell off the deck of an aircraft carrier in September 1976, not to mention the cost of recovering it.

But whether the money comes from a shortfall or a new spending priority, it is imperative that this nation, at this time, pledge the resources that must be committed to this task. It is in our own national interest to do so; that is, if we see the alleviation of human suffering as a national goal.

I cannot urge more strongly that the new administration place on its foreign policy agenda the increased support and funding of this nation's worldwide activities in famine prevention. Such a step would represent a dramatic shift in our foreign policy emphasis. It would signal to the poorest of the poor in the developing world that the United States believes freedom from hunger is a more important international goal than the sale of arms. It would link our foreign policy with the self-help and humanitarianism of mainstream American values. And above all, it would marshal the necessary resources and human talent that could, once and for all, end the scourges of malnutrition, hunger, famine, and death by starvation.

We have shown the way in our own country. In 1870, the average farmer in the United States produced enough food and fiber for about five people. By 1930, farmers had doubled their output to support 10; by 1955, production had again doubled so that the farmer was feeding 20 people. Today, the same farmer has again more than doubled his output, producing enough to feed himself and 51 other people.

We can be proud of this record of achievement—one of the modern miracles of the application of science and technology. The challenge for the next 25 years is to assure similar growths in agricultural productivity, both at home and abroad.

THE ECOLOGICAL ISSUE

As we unleash with renewed vigor the power of our vaunted scientific capabilities, we must be mindful of the balance which must be found between ecological considerations and agricultural development.

On the one hand, we have a science of agriculture that has brought humankind great success by its skillful and productive interventions in nature for man's benefit. But on the other hand, this same science is telling us more and more about the often-dismaying consequences of our interventions. Thus, the irony: the Malthusian specter requires us to produce more and more food to keep up with rising population plus demands for better nutrition worldwide at the very time that we are reaching a proper awareness of the effects of our massive interventions in the ecological balances of the world.

We find ourselves with deeply held values that are in conflict. The humanitarian in all of us drives us to produce more and better food for the starving and undernourished family of man. But at the same time, as good stewards of the land, we are driven to protect and preserve the life-sustaining potential of our planet Earth.

We are painfully aware that more extensive and intensive agriculture often endangers species of animals by destroying their habitats or introducing toxic chemicals into their environment; that it often endangers fresh water supplies by the runoff of fertilizers, threatening lakes with eutrophication and fish sources with extinction; and that it often endangers the climate itself by the overuse of fragile land which can turn into dust bowls, thus increasing the particulate matter in the atmosphere, reducing solar energy, and affecting the world's temperature.

As we look to the future, we need some new style of response, I believe, to balance the claims of continued growth against those of ecology. What we need is a reexamination of our philosophical approach—the humane perspective which determines the way in which we deal with these problems and their solutions.

The last chapter of the 1975 *Yearbook of Agriculture* is entitled "The People-Food Race, and How to Win It." It is written by four food experts, Joseph J. Marks, H. R. Fortmann, J. B. Kendrick, and Sylvan H. Wittwer (1975). The four coauthors believe that agricultural problems are no longer the exclusive property of agricultural scientists.

The most important thing to realize now, they say, is a "feeling that we must work together for the whole human race." ". . .That caring for all those who reside on this humble planet. . .is the most important influence on research priorities. Our research programs must know no borders, geographical or otherwise. We must avoid being locked into old formulas, organizational patterns and concepts." As we build that international resource we call food production, they conclude, we must also build up our new intellectual resource which they call "environmental management."

I fully agree with their conclusion. We must maintain a delicate balance between the urgent claims of humankind for food now and the long-range claims of the environment with its finite resources of energy, land, water, and nonrenewable materials.

The goal must be no less than that of developing a new style of environmental management, the management of interdependence, based upon our new intellectual and ethical understanding of systems. Such a new style would be worthy of being called "husbandry"—in the best traditional sense of the term—a choosing, balancing, allocating, and managing of resources within a deeply-felt commitment to the interdependence and interrelationship of man with nature and man with man. We must strike a human balance. If we succeed in developing this more human ecology, we will, I believe, make ourselves more humane.

Our commitment must be the production of food for the starving, the malnourished millions, now alive and soon to be born. But our commitment must also be the preservation of the food-producing, the life-sustaining potential of the planet Earth. Not merely "departmentally" are we the food producers; we are also the stewards, the husbandsmen of global fertility, the managers of fragile ecosystems which are the basis of life for all mankind. Our relationship to that system, as well as to the multitudes of mankind, must be part of that balancing of considerations that goes into our management of the world's resources. Within a new sense of intellectual and ethical interdependence, we hold both the present and the future in trust.

Ours, I believe, is the mission of "the good steward," and our estate is nothing less than Earth itself. To us is entrusted the task of husbanding the earth's productive capacity into the long-range future, for the family to whom and for whom we are ultimately responsible—the family of man.

CONCLUSION

One of the most puzzling arguments of the neo-Malthusian prophets of doom has been their contention that man somehow has reached the limits of technology, that scientific and technological improvement of agriculture and food production is somehow artificial, and that reliance upon such methods has been an evasion of nature and environmental law. I have no idea why intelligent human purpose should be considered "unnatural." The mind of man is as much a part of nature as the fertility of the soil, or, for that matter, the tug of gravity. Far from having reached the limits of science and technology, I believe that we have only begun to explore our capacities for increasing agricultural productivity through basic and applied research.

Herein lies the potential for unlocking the "Malthusian prison."

The creative partnership of the universities' research and education capacity with farmers and agricultural business enterprises has been the key to the marvel that is U. S. agricultural productivity. Title XII offers an unprecedented opportunity to test, extend, and apply our resources and talents to the agriculture upon which three-fourths of mankind depends. By recognizing the crticial problems, by striving to understand their complexity, by cooperating to provide inventive and dynamic solutions, and by striving to achieve a human balance, we can avert disaster and for the first time in the history of civilization adequately feed the peoples of the earth.

Meeting the world food challenge also requires both commitment and faith. Our commitment must be to the resolution of the problem. Our faith must be in man's infinite creativity.

Although creativity is hard to predict and quantify, history bears out its existence. Just as Malthus could not foresee chemical fertilizers, the modern tractor, hybrid corn, or any of the countless innovations that have helped create modern agriculture, we cannot today predict the developments which will take place in our own future. But we can make one forecast with great confidence: innovation and new discoveries will continue to take place as long as thinking man exists.

While we must not offer false reassurances on the immediate and magical properties of science and technology, past achievements provide dramatic evidence from which to predict long-run success. We must also recognize the other important elements in the world food drama—population growth, socioeconomic institutions, and political conditions. But the role of science and technology remains basic.

Advances in science and technology have not been without their own perils or unanticipated consequences, as environmentalists are quick to charge. But if we can strike a human balance, I think that the assets deriving from science and technology outweight the liabilities. Basic and applied research must continue to grow, for they are the catalysts that push ever upward the quality of life. To be ready with solutions to the problems of tomorrow, we must be willing to allocate economic and manpower resources today. If we wait until the future is upon us to react to it, we shall have failed already.

I have no quarrel with the doomsayers over their projections or their facts. My concern is that their warnings not become the rationale for inaction. Their counsel should not be of despair but a call for even greater efforts to meet, and hopefully overcome, the threatening disaster.

The worldwide food crisis is real. It is serious, and it will probably worsen. Worldwide famine in all probability will be a recurring theme in our lifetime. But it is not inescapable, provided we hold fast to the following article of faith: While the world's resources may be limited, we have yet to discover the bounds of human creativity.

LITERATURE CITED

Epstein, E. J. 1976. Good news from Mr. Bad News. New York Magazine 9(Aug.):34–39, 42–44.

Marks, J. J., H. R. Fortmann, J. B. Kendrick, and S. H. Wittwer. 1975. The people-food race, and how to win it. p. 345-350. *In* 1975 Yearbook of Agriculture. USDA, U. S. Government Printing Office, Washington, D. C.

Meadows, Donella, D. L. Meadows, Jorgen Randers, and W. W. Behrens, III. 1972. The limits to growth. Potomac Associates, Universe Books, New York, N. Y.

Agronomists and Food—Contributions[1]

D. Wynne Thorne

Agronomists have been major contributors to the supply, diversity, and quality of food. Identification of these contributions over the life of this nation depends, however, on the definition of an agronomist.

The first president of the American Society of Agronomy, M. A. Carleton, defined an agronomist as a person concerned with the science and art of crop production and soil management. This definition is broad enough to include engineers, entomologists, and those of many other disciplines. But today agronomists must extend their horizons even further to be concerned with quality of environment and the conservation, protection, and wise use of our natural resources. For our purpose, the broader concept of crop production, soil management, and conservation of natural resources with high regard for environmental quality seems to best encompass the scope of agronomic contributions.

The cooperative endeavors of scientists from many fields have been the key to major advances in food production and the integrating role of agronomists in such scientific teams is of paramount importance. I will list here advancements in which many scientists have played an important part. To a number of specialists in such diverse fields as engineering, entomology, plant pathology, and plant physiology, as well as agronomy, I am indebted for suggesting many more scientific contributions than could be included in this paper.

D. Wynne Thorne is Professor Emeritus of Soil Science and Biometeorology at Utah State University, Logan, Utah.

[1]Contribution from the Department of Soil Science and Biometeorology, Utah Agricultural Experiment Station, Logan, Utah. Journal Paper No. 2120.

THE FIRST ONE HUNDRED AND TWENTY-FIVE YEARS

Since my theme is tuned to the Bicentennial, let us look at farming 200 years ago. In many aspects we could be looking backward 2,000 years with only minor differences. The writings of the period leave little doubt that the knowledge and practices of agriculture in the late 1700's had advanced very little since Roman times.

One exception is the wealth of important new crop plants obtained from the American Indians. Early settlers brought seeds of many European crops with them, but they often had better success with crops such as maize (*Zea mays* L.), pumpkins (*Cucurbita pepo*), squash (*Cucurbita maxima*), peas (*Pisum sativum* L.), and beans (*Phaseolus vulgaris* L.) adopted from the Indians. From the Indians they also obtained tobacco (*Nicotiana tobacum* L.), white potatoes (*Solanum tuberosum* L.), sweet potatoes (*Ipomoea batatas* Lam.), tomatoes (*Lycopersicon esculentum* spp.), green peppers (*Capsicum* sp.), and sunflowers (*Helianthus annus* L.).

Corn and pumpkins were especially significant to the early American colonists. Maize was widely adapted, had many uses, provided food for both animals and humans, and stored well. Pumpkins were used so extensively that when the Pilgrims commemorated Thanksgiving Day, it was often referred to by their neighbors as St. Pumpkin's Day.

In 1776 most Americans owned a farm and two out of five made their living entirely by farming. According to one report, two men and two to three horses or four to six oxen were required to plow one-half to one hectare per day. One man could handle about two hectares of plowland. Crop production practices included plowing in the spring, planting of crops with little or no cultivation (weeds grew in competition), and obtaining outside help for harvesting. Scythes or sickles were used for harvesting grains. A small farm might have consisted of 40 ha (99 acres) with 2 ha (5 acres) farmed intensively, 12 ha (30 acres) superficially, 1 ha (2.5 acres) or 2 ha (5 acres) of orchard, and most of the remainder used for pasture. The average yield of wheat was reported as about 650 kg/ha (580 lb/acre) and corn as 900 kg/ha (803 lb/acre) to 1000 kg/ha (892 lb/acre) (Carrier, 1923).

The 30 years from 1820 to 1850 were described by William Jardine (Secretary of Agriculture 1925–1929) as "the beginning of a fabulous era in agriculture" because of the many important farm machines developed. Some of these machines that changed farming practices included: the horse drawn revolving hay rake, the reaper, the thresher, the combine, the steel plow, the grain drill, the horse drawn wheel cultivator, the disk plow, and the first portable steam engine. Jardine stated of this period, "The methods of crop production underwent greater changes than they had in the previous 5,000 years. At one stride we covered ground where 50 centuries had left almost no mark." The names of John Nuebold, Jethro Wood, John Deere, and Cyrus McCormick stand out as pioneering inventors of these machines (Promsberger, 1976).

As machines were substituted for human labor, the capability for farming larger areas of land was greatly increased, but crop yields were not appreciably affected. By 1866 when crop statistics became available, the average yield of wheat (*Triticum aestivum* L.) was 650 kg/ha (580 lb/acre), the same as reported nearly 100 years earlier. Corn, however, was averaging about 1400 kg/ha (1249 lb/acre) compared with 700 kg/ha (624 lb/acre) to 900 kg/ha (803 lb/acre) in the pre-revolutionary period.

The Search for Knowledge

Not until about 1840 did man begin an intensive search for knowledge about soils and plants. The curiosity and studies of such European scientists as G. F. Liebig, Francis Home, and J. B. Boussingault gradually produced laboratory, greenhouse, and field experimental methods. Evidence was gradually obtained that plants need substantial quantities of the ten major essential elements. Edmund Ruffin of Virginia published his studies on calcareous manures in about 1845, giving new evidence of benefits from the ancient Roman practice of adding lime to soils. J. T. Way of England opened a new field of thought and investigation with his discoveries that the clay fraction of soils has the capacity to adsorb and release cations (Russell, 1973).

These advancements in knowledge of plant nutrient needs and of their relationships to soils fostered the beginnings of fertilizer development and use. Most important was the first production of superphosphate by the treatment of bones with sulfuric acid by J. B. Lawes and S. G. Gilbert, founders of the Rothamsted Experiment Station in England. Others found benefits to plants from treating soils with potassium and nitrate containing salts. But supplies of such substances were limited, not easily handled, and relatively expensive, thus causing delays in their general use.

Just before 1776, a series of experiments involving J. Priestly, J. Ingen-Housz, Jean Senebier, T. de Saussure and others showed that, in the presence of light, plants absorb CO_2 and give off oxygen, while in the dark they behave like animals and take in oxygen and give off CO_2. But, except for Liebig's use in arguments against the traditional belief that plants obtain organic substances from soil humus, this information was largely ignored by plant scientists until the 1860's. Then studies on photosynthesis started.

About the end of this nation's first century, increasing attention was focused on soil microorganisms. The most practical applications of these investigations related to soil organic matter and nitrogen transformations. Organic matter was shown to be a dynamic part of the soil, serving as a reservoir and supplier of plant nutrients and as a soil conditioner.

Many organisms capable of fixing atmospheric nitrogen were identified and studied, but the most important discovery relative to improved food production was the symbiotic fixation of nitrogen. Evidence for this process included the work of H. Hellriegal and H. Wilfarth in Germany in demonstrating that legumes can augment available nitrogen in soil and sand cultures, and of M. Beijerinck who isolated and identified the bacteria in the root nodules.

It was not until the 1890's and early 1900's that soil microbiology gained momentum in the United States. One important area of discovery was of the many cross-inoculation groups of legume bacteria. Later, centers such as the Fixed Nitrogen Laboratory at the University of Wisconsin, probed the fundamental reactions in the nitrogen fixation process. In recent years there has been an international rejuvination of interest in the entire process of biological nitrogen fixation.

With its great importance, soil water naturally received considerable attention and various systems of describing and classifying soil moisture were proposed. It was not until the studies of G. Buckingham in 1907, followed by those of Willard Gardner which started in 1920, that the concepts of energy and work in soil moisture movement and absorption by plants were placed on a sound basis.

The release in 1900 of information about G. Mendel's earlier discoveries stimulated plant scientists to envision new and improved crop varieties. In 1903, W. M. Hays of the University of Minnesota founded, and became the permanent secretary of, the American Breeders Association. Others involved included H. J. Webber of the U. S. Department of Agriculture (USDA) who was concerned with hybridization experiments with cotton and citrus; W. J. Spillman who was starting a wheat breeding program at Washington State College; E. M. East of the Connecticut Experiment Station who became interested in corn; R. A. Emerson of the University of Nebraska who worked with East on corn genetics; and G. H. Shull of the Carnegie Institute, also concerned with corn breeding.

From this group and their associates and students, rose much of the early plant breeding research in the United States. The American Breeders Association terminated in 1909, no doubt in part as a result of the organization of the American Society of Agronomy in 1908. The original group, however, had provided the foundation for a rapidly evolving science of genetics and its application in successful plant breeding programs (Warburton, 1933; Hays, 1951).

Of significance in the first one and a quarter centuries were the organization of the U. S. Department of Agriculture, the federal authorization of land grants for colleges of agriculture in 1862, and the Hatch Act of 1887, providing support for an agricultural experiment station in every state. In 1914 the Smith Lever Act provided support to states to establish extension programs to carry research findings to farmers.

Thus we see that the 19th century and the beginning of the 20th were largely concerned with gathering knowledge about soils and crop plants, testing various crops and crop varieties, and evaluating various soil management and tillage practices. Some of the more readily applied products of research, such as identifying the higher yielding strains of plants, were adopted by farmers. But to a large degree, the scientists were regarded by the farmers as impractical dilettantes. This was, however, a period of great expansion of cultivated land and of regionalization of crop production. With the westward expansion of agriculture, the corn belt, the wheat belt, irrigation general farming, and forage and livestock regions gradually emerged. Also, the old cotton belt expanded westward.

THE TWENTIETH CENTURY

Soil Management and Fertility

The rapid westward movement of agriculture in the 1800's had involved the plowing and cropping of the better and most easily available soils. The native fertility of the soils was mined and even elementary conservation practices were commonly ignored. The early 1900's reaped the inevitable consequences—depleted soils, increased and new types of nutrient deficiencies, erosion of the topsoil, severe dust storms, and increased damage to crops from insects and diseases. The evidence clearly showed that a permanent and prosperous agriculture and an assured supply of food could not come from continually plowing more land. Farmers must learn to husband their resources, not exploit them.

Men with vision saw that the investigations begun in the latter part of the 1800's and the early 1900's must be intensified and directed toward solutions of specific problems. The findings must be taken to the farmers and presented in such a way as to assure their adoption. Larger numbers of trained scientists must be produced.

During the first half of the 20th century the list of essential elements was expanded from 10 to about 20, with most of the investigations now being carried out in the United States. Strong research teams were developed in Florida with O. C. Bryan and others, in Kentucky under the leadership of J. S. McHargue, and in California involving A. R. C. Haas, D. R. Hoagland, P. R. Stout and others. Workers in Florida and Australia added greatly to knowledge about practical field use of the essential minor nutrient elements (Tisdale and Nelson, 1975).

The great advance in nitrogen fertilizer use came from the discovery of the process of synthetic fixation of nitrogen by F. Haber and C. Bosch in Germany, with the first plant going into production in September of 1913. During the past 50 years the investigations of USDA and later of TVA personnel have made the United States the leader in fertilizer technology with many new types of fertilizer being developed as a result of TVA studies.

Numerous studies on fertilizer use have resulted in specific improved technologies. Major advances have included development of soil tests to determine the quantities of specific nutrients available in soils (Emil Truog, R. H. Bray, G. N. Hofer, S. F. Thornton, R. K. Scholfield, Sterling Olsen).

Extensive investigations have resulted in more efficient uses of fertilizers through adjusting rate, time, and place of application. Studies such as those of George Scarseth, A. J. Ohlrogge and B. A. Krantz, with varying combinations of fertilizer, varieties, and plant population, laid the foundation of increased rates of application of nitrogen fertilizers that have been so important to increased corn yields.

Soil Improvement

High among the scientific findings providing a base for productive soil management was the discovery of the nature of clay. R. Bradfield and others had shown clay to be minerals rather than amorphous colloids. In 1930 S. B. Hendricks and W. H. Fry of the USDA, using x-ray diffraction patterns, revealed the crystal structure of clays. This was followed in 1931 by a similar and verifying paper by W. P. Kelley, W. H. Dore and S. M. Brown of the California Agricultural Experiment Station (Kelley, 1940). These papers provided a background for the understanding of cation exchange and of differences in the swelling and contracting of soils. W. P. Kelley and others used these findings to develop concepts and practices for reclaiming and managing saline and sodic soils.

In the more humid regions, related clay mineral research has helped define the role of exchangeable aluminum in acid soils. Aluminum toxicity to plants and the role of lime in its control are basic to understanding and managing leached and acid soils.

Soil depletion and erosion have been common adjuncts of farming through the ages, but such problems were accelerated with tractor power and farm mechanization, particularly as cultivated crops invaded increasingly arid zones such as the Great Plains. M. F. Miller and H. H. Krusekopf of the University of Missouri initiated the first field plots to evaluate factors affecting runoff and erosion. In 1929 Congress provided funds to establish ten field stations to study erosion in cooperation with various states. A few years later the Soil Erosion Service was established under the leadership of Hugh Hammond Bennett. Bennett was a vigorous apostle and spokesman for the soil conservation movement and was the important leader in developing the Soil Conservation Service.

Although the data from the soil erosion field plots is best known through its use to emphasize the importance of various soil conservation practices, probably the most significant application is that of the "Universal-Soil-Loss equation." W. H. Wischmeier and D. D. Smith of USDA, located at Purdue, were asked to analyze the available data on soil erosion and conservation practices. The equation they produced has become a standard reference and tool for planning and evaluating soil erosion problems and corrective practices.

Various conservation practices and specialized instruments have evolved from the many studies. Among them were the stubble mulch practices developed by F. L. Duley and J. C. Russel in Nebraska and various tillage implements such as the Noble Blade and the Graham Hoeme Plow which left crop residues on the soil surface and the soil in condition to resist wind and absorb rain (McCalla and Army, 1961).

Important to conservation and the entire problem of inventorying soil resources and transferring management technology was the development of a modern scientific system of soil classification. This was initiated by C. F. Marbut based on concepts borrowed from V. V. Dokuchaiev and others, and developed into a modern world system under the leadership of Charles E. Kellogg and Guy D. Smith.

One of the major advances in this century of farming has been the development of integrated soil conservation and farm production systems tailored to the resources and needs of individual farms. This program of farm planning as conceived and developed by the Soil Conservation Service stands as one of the notable achievements in the application of science to farming. Agronomists assumed a major role in this program. The integration of principles and practices at the farm level is a final and essential link in maintaining a viable agricultural production system.

Plant Breeding

Plant introductions have been an important part of crop improvement in this country. Such genetic imports as dwarf wheat from Japan, soybean collections from Asia, crested wheat grass from Russia and bermudagrass (*Cynodon dactylon* L.) from Kenya and South Africa are examples of recent introductions that have played important roles in plant breeding programs.

Great advances in plant breeding were first realized with corn and wheat. These are well known and will not be elaborated here. Soybeans were of little interest until intense research produced new varieties such as Lincoln. Soybean (*Glycine max* L. Merr.) breeding has featured improvements in photoperiodism, resistance to phytophthora rot and the cyst nematode, improved resistance to lodging, and improvement in yield potential and oil and protein quality and content. The breeding program has been centered at state experiment stations in close cooperation with the USDA.

Alfalfa (*Medicago sativa* L.) is an early crop introduction that first gained popularity in California and other western states. Variety introductions and breeding have produced improved winter hardiness, and resistance to bacterial wilt, phytophthora, stem and root nematodes, and several insects. These research achievements have made alfalfa the queen of forage crops.

In grass breeding the principal success has been the production of Coastal bermudagrass by hybridizing Georgia and South Africa strains of bermudagrass. This variety with its superior yield and palatability was the joint effort of USDA and Georgia Experiment Station scientists led by Glenn Burton.

The sugar beet (*Beta vulgaris* L.) saga centers around cooperation between plant breeders and mechanization specialists. The breeders had produced resistance to curly top virus and other diseases and insects, but the labor of thinning and harvesting the crop were production barriers. A seed ball containing a single seed was needed. A report was received that a Russian refugee, then in Europe, knew of such a genetic source. V. F. Sevitsky and wife were brought to the United States and worked in Utah and California. The secret was that the early bolting trait was closely associated with monogerm seed. With this information, monogerm seed was located and bred into improved beet varieties. This, combined with precision drills and weed control, solved the problem of securing well-spaced single plants with a significant reduction in hand labor.

After E. T. Mertz and others at Purdue discovered a high lysine maize associated with an opaque endosperm, other intensive searches for cereal germ plasms having high contents of protein or unusual proportions of specific amino acids were begun. New discoveries have been reported for sorghum by J. D. Axtell at Purdue and for barley by L. Munck and others at the Swedish Seed Association (Brown et al., 1976). The urgency for higher protein foods has been moderated by the findings of nutritionists that many reported protein deficiency symptoms in developing countries are actually a result of calorie deficiencies. Despite the recent lowering of the minimum protein requirements, the new improved protein lines coming out have a great potential for improving human and animal nutritional levels.

Water

In terms of quantities required and the rapidity and drama of deficiency symptoms, water is the most significant of all factors required for plant growth and thus for food production. The development of the soil tensiometer, as conceived by L. A. Richards and W. Gardner in 1936, to measure the capillary potential of soils, marked the beginning of intensive technology development for measuring and controlling soil moisture for plants.

F. J. Veihmeyer and A. H. Hendrickson, beginning in 1924, pioneered studies of irrigation practices in relation to yields of fruits and other crops. Their conclusions, that fruit yields were broadly the same over the soil moisture range between soil field capacity and the permanent wilting percentage, stimulated much research. In 1952, L. A. Richards and C. H. Wadleigh published an extensive review of accumulated data, providing conclusive evidence that the growth of plants decreases as soil moisture stress increases over most of the available field moisture range.

Extensive studies have explored the water requirements of crops in relation to soils and climate along lines conceived by L. J. Briggs and H. L. Shantz in Arizona in 1913. These studies have shown that high yielding crops consume only a little more water than low yielding ones and that, for certain cropping conditions, the water required for crop production can be predicted from climatic variables. One result has been the advantageous use of climatic data in planning water needs in new irrigation projects. Additionally, it was realized that the most effective way to attain maximum water economy is by combining efficient irrigation practices and the use of fertilizer and other inputs to maximize yields.

Since it is difficult to maintain soil moisture near optimum without excessive losses, many changes have been made in irrigation practices and technologies from wild flooding to furrows, borders and basins, to diverse types of sprinkler systems. Recently, use of trickle irrigation has increased substantially for fruits and other high return crops. Investment in such systems is large, but water economy, savings in labor, and the capability of sustaining high crop yields has resulted in a continual shift to more controlled irrigation systems.

Important to efficient irrigation are procedures for indicating when and how much water to apply. The porous plate apparatus developed by L. A. Richards has made practical rapid and reasonably accurate estimates of field capacity and the permanent wilting percentage. Also, the precision has been aided by the development of irrrigation schedules based on estimates of evapotranspiration through such formulas as those developed by H. F. Blaney and W. D. Criddle and M. E. Jensen. These, combined with such soil moisture measuring devices as the tensiometer, the gypsum block conductivity apparatus of George Bouyoucos, and the more recent neutron probe have taken much of the guess out of water application (Haise and Hagan, 1967).

Important to a permanently successful irrigation agriculture is drainage and control of salt in soil. Starting with E. W. Hilgard (1906) and extending through the studies of W. P. Kelley and many others in various agricultural experiment stations of the western states and of the U. S. Regional Salinity Laboratory, precise methods for evaluating and controlling salinity and sodium accumulation in soils have been developed.

The acreage of irrigated lands is continually increasing, accounting for over 10% of all cultivated lands and an even greater proportion of the total production of food. But in rain-fed lands the principles of moisture conservation and effective use through adjusting crop planting dates, quantities of fertilizer, plant population, slopes and shapes of rows, and tillage practices have received increased worldwide attention and are now important items in integrated farming systems. Minimum and no-tillage practices are recent conservation developments based on the combined use of herbicides and special planting and tilling equipment (Baeumer and Bakermans, 1973).

The Age of Chemicals

Weeds, insects, and diseases have been enemies of the farmer throughout time. Until recently, for much of the crop growing season, there was a continuing contest between the crop and weeds and insects.

The major advance in weed control was the discovery of new herbicides during the screening of many chemicals for biological properties during World War II. Apparently 2,4-D was discovered independently in England and the United States in the early 1940's. This and a number of other selective herbicides were identified in the Special Projects Division of the Chemical Warfare Service at Camp Detrick, Md. A large number of agronomists participated in this program. Within five years after its discovery, 2,4-D was being used to control weeds on more than 18 million acres of small grains and on 4.5 million acres of corn in this country alone, an unprecedentedly rapid adoption of a new technology (Ennis and McClellon, 1964).

Development of new insecticides essentially paralleled that of herbicides. In 1939 the potato crop in Switzerland was threatened by the Colorado potato beetle. The Geigy Chemical Company made samples of DDT available for testing on the beetles. Results were so spectacular that trials were soon extended to other insects and crops. Dr. J. Paul Müller of the J. R. Geigy Company of Basel, Switzerland was awarded the 1948 Nobel Prize for development of DDT. Some have said the major discoveries in curative and preventative medicine during the war were DDT, plasma, and penicillin. Since then the discovery of synthesis and testing of many insecticides and herbicides have become routine.

This fantastic chemical age has also included the discovery of growth stimulating substances in plants such as auxin and β-indolacetic acid, gibberellin, and other hormones that affect plant development, fruit set, and size.

Add to this the growing use of chemical fertilizers, and crop production had become an integral part of our chemical age. For a short period this seemed like manna from heaven, reducing human bondage to burdensome tasks and providing solutions to age-old problems. But this new fruit was found to contain bitter spots. Excessive or careless use of fertilizer polluted streams and lakes; some insecticides harmed birds and other forms of animal life and killed desirable as well as harmful insects. Residual effects of herbicides sometimes harmed subsequent crops planted on treated areas. Carcinogenic properties were attributed to residues of some agricultural chemicals. And to add to the dilemma, with the energy crisis came the sober constraint that chemicals may be too costly in terms of energy for many uses. Frequently the opposition to chemical uses became unreasonable and threatened restrictions that would either endanger the supply or greatly increase the cost of food with questionable benefits.

In the face of this growing crisis many agronomists stood up and spoke out clearly in the midst of a national environmental pandemonium. Such men as Samuel Aldrich of the University of Illinois advanced sound testimonial evidence on the place of agricultural chemicals in environmental situations where hysteria had threatened to reign. Also the Council for Agricultural Science and Technology (CAST) study groups organized under the leadership of Charles Black and others have provided rational, professional advice on controversial issues.

Systems of management have gradually evolved in which each agricultural chemical is no longer viewed as a specific solution to a particular problem. Rather, farming programs are planned as an entity. The chemicals used must fit into tillage practices, crop sequences, crop varieties, and soil management programs to create integrated systems. In this complex program the agronomist has a central role, a role that requires different training and a comprehensive vision of the interrelationships of many complex variables. Increasingly, agronomists must function as a part of, and often a central figure in, interdisciplinary teams of scientists.

Integrated Production Practices

The disciplinary advances in soil management, irrigation, plant nutrition, and fertilizer technology; the control of pests; and the breeding of superior plants have each exerted notable advances in food production. But the major boosts have come from optimum combinations of these factors.

Such integration of production factors began under irrigated conditions in the West and in the favorable climatic situations of the Corn Belt. The undergirding data came from investigations of combinations of improved crop varieties, various levels of fertilizer nutrients, adjustments in plant population, and, under irrigation, frequency of water application. The results demonstrated dramatic increases in yields as optimum combinations of production factors were brought together. Pest control has been a difficult variable to include in such experiments, but its significance in improving production under farm conditions is unquestioned. Great numbers of experiments involving diverse crops and soils and conducted by numerous investigators, have led to the formulation of principles that are now accepted as the basis of modern agriculture. This integration of production has been the crowning achievement in the contribution of agronomists to world food supply. Perfected largely in the United States, the technologies and principles have been exported to developing countries as the basis of the green revolution.

What has this meant to food supplies? We can examine the yields of our crops in the middle of the 1940's and compare them with yields in recent years. Average yields of corn, grain sorghum [*Sorghum bicolor* (L.) Moench.], potatoes and tomatoes have trebled. Wheat and barley (*Hordeum vulgare* L.) yields have doubled, and accompanying these soaring yields have been improved quality and increased uniformity of harvested products.

Education

The education of agronomists has been an essential and leading factor in improving food production. It is the power that makes our system of scientific agriculture move forward. The Land Grant Act signed by Abraham Lincoln in 1862 provided the state and national system of agricultural colleges. Undergraduate and graduate degrees in agricultural sciences have increased consistently, reaching 10,000 Bachelor of Science degrees per year and 18,000 students registered in graduate programs in the early 1970's.

The heart of education is always the great teachers, and among these agronomy and related subjects have had their full share. A few are listed as examples of a large and insufficiently recognized group.

H. K. Hayes of the University of Minnesota is often referred to as the father of plant breeding in America. He is remembered by his students for his integrity and high intellectual standards. He was an industrious worker who demanded excellence of performance in his associates and students. A large portion of our leading plant breeders obtained their training and inspiration from him.

A philosophic, friendly relationship with students was the trademark of William Albrecht of the University of Missouri. He attracted large numbers of students and inspired many to succeed as soil scientists.

As I have visited cereal experimental plots throughout the Near East, I have found a large and growing group of young plant breeders and production agronomists who look with reverence to Norman Borlaug, of the International Maize and Wheat Improvement Center (CIMMYT) in Mexico, as their teacher. One of their common questions was "Do you think Dr. Borlaug would like what I am doing?"

The contagious enthusiasm of Thomas L. Martin of Brigham Young University attracted students to agronomy from many other disciplines and inspired them with the challenge of excelling as professional agronomists. F. D. Keim played a similar role at the University of Nebraska. Henry Smith of Washington State University set high standards of excellence that stimulated improved career performance by many students.

The list could be expanded. In recent years numerous teachers have responded imaginatively to the challenge and demands of students with new approaches to instruction and even higher standards of teaching.

Extension

Preceding and since the Smith Lever Act of 1914, extension agronomists have been in the forefront of educating U. S. farmers. Through the efforts of these extension specialists, farmers have come to know of new varieties and improved production practices. Over the years they have demonstrated new technologies so effectively that the gap between discovery and practice has almost disappeared. There can be no clear separation of the returns from extension and from agricultural research, because extension in its application is the demonstration and adoption of research technology by the public.

About 40% of all U. S. extension workers are concerned with improving agricultural practices. Some examples of close cooperation between extension and research include annual guides issued by extension services in most states for farmers, providing directions for the use of herbicides, insecticides, fungicides, and fertilizers. Tests and demonstrations are conducted in farmers' fields on new crop varieties in almost every county. Demonstrations and illustrated slides are prepared and made available on most new farm practices such as no-tillage and minimum tillage crop management systems. A further important joint service has been the development in each state of a seed certification program to provide sources of pure seed and assure seed purity in the seed trade.

International

With the rapid strengthening of agricultural science and agricultural productivity during and following World War I, and especially following World War II, other nations looked increasingly to the United States for advice, training, and technology in agriculture.

Among the early agronomists to respond to the challenge was Cyril G. Hopkins of the University of Illinois. He, with George Bouyoucos of Michigan State University, went to Greece in the fall of 1918 to advise on increasing food production. Enroute home after a successful year, Dr. Hopkins contracted malaria and died at Gibraltar.

Some ten years before Hopkins' mission, Franklin H. King of the University of Wisconsin had gone to Japan and China to study how farms had remained productive after several millenia of continuous farming. His book, *Farmers for Forty Centuries* has had a major impact on the thinking of agronomists and others on the potential longevity of farming.

Numerous success stories attest to the contributions of agronomists in augmenting food production in developing countries. In Kenya, U. S. sponsored programs made hybrid corn seed available to farmers with the limitation that they would also use recommended packages of improved practices. Hybrids were used as a lever to speed the use of combinations of recommended programs. The results are striking. The demonstration farms yielded about 3,500 kg/ha (3,122 lb/acre) while nearby farms were producing 800 kg/ha (714 lb/acre) to 1,500 kg/ha (1,338 lb/acre). As a conservative estimate, 300,000 ha (741,000 acres) were planted under the new practices within two years, providing an increase in crop value of $7 million.

Four U. S. extension advisors helped train 1,000 Vietnamese extension workers in using 19 clearly defined steps for growing improved crop varieties. These extension workers, in turn, contacted and helped 300,000 farmers who tripled rice yields in much of the country.

It is difficult to measure the impact of the numerous technical programs that have been carried out in diverse developing countries. Many contracts have been of too short duration to accomplish major goals. Many were relatively ineffective because of short range U. S. government policies and administrative interference. Nevertheless, the total impact has been great. One important aspect has been the training of nationals for nearly every country involved. Much of the advanced training has been in the United States, but this has been bolstered by on the job guidance by agronomic specialists as students returned from the various U. S. colleges. Facilities have been provided and national research programs initiated in these countries with agronomists in the forefront in helping to develop combinations of improved technologies suited to local conditions.

Today the technical assistance limelight is focused on the international institutes and centers. Twelve such centers have been approved and most of them have well-developed programs. These centers are now financed and administered through the Consultative Group in International Agricultural Research. Participants include about 30 nations, 3 foundations, the World Bank, and agencies of the United Nations. The United States currently provides about one-fourth of the core support for each of the centers.

The longest operated and best known of the institutes are the International Rice Research Institute (IRRI) in the Phillipines and CIMMYT in Mexico. These two were established by the Rockefeller and Ford Foundations and later taken over by the Consultative Group.

The role of these centers as a basic force behind the "Green Revolution" is so widely accepted as not to need retelling. CIMMYT was initially sponsored by the Rockefeller Foundation. In 1943, George Harrar became the first leader. Edwin Wellhausen was selected to lead the maize program; Norman Borlaug, wheat; and William Colwell, the soils and farm practices program. The Center has produced dramatic results with maize and wheat in Mexico. In 1973, for example, the Center furnished seed stocks to 93 different nations. Many Mexican nationals have been trained and other trainees were attracted from many countries. In less than 25 years after the start in 1943, corn production in Mexico had trebled. Average yields increased from 500 to 900 kg/ha (448 to 784 lb/acre). Wheat, grown primarily on irrigated land, made even more spectacular achievements. Short-strawed, photo-insensitive, and disease resistant Mexipack wheats produced have become the base for improved wheat yields around the world. In Mexico, wheat yields averaged 770 kg/ha (687 lb/acre) in 1943, and in 1963 yields were 3300 kg/ha (2944 lb/acre) to 5900 kg/ha (5263 lb/acre) in the major production areas. Mexico has become an exporter rather than an importer of wheat (Stakman et al., 1967).

Similar achievements can be reported for several International Centers. The programs of others are just gaining momentum, but the prospects are bright.

Where Are We?

Our survey of the past 200 or more years have shown that knowledge, technology and practice in food production in the United States have advanced in quantum increments. Some of the periods and advances since 1776 have included the following:

1) The first 50 or more years was primarily an exploitation of this nation's soil and other natural resources and trials of various crops under different environmental conditions.
2) From 1820 to 1850 came the mechanical revolution when many of our most important farm machines were invented. Man's capacity to exploit natural resources increased; more land was plowed but crop yields remained low. The basis and processes of soil depletion and soil erosion and the dust bowl were growing.
3) About 1840, the vigorous search for knowledge about plants, soils,

water, and the principles of crop and food production began. This has continued to the present, but it was not until well into the 20th century that this accumulated knowledge and technology produced the rewards in terms of improved farm production practices. Education was an important and integral part of this total program.

4) Shortly after 1900 came the advances of genetics and plant breeding that produced an important basis for a revolution in farming in the 1930's.
5) In the 1920's combinations of education, research, and extension were attained to provide a basis for transferring knowledge and technology to users.
6) In the 1930's came drought and the dust bowl with a vigorous national response to the challenge. The Soil Conservation Service was established and agronomists helped develop and put into practice sound soil and water conservation practices.
7) About 1945 the beginning of the chemical age burst upon agriculture and food production. DDT, 2,4-D, the vast expansion of fertilizer use, plant hormones, growth regulators, and the integration of these into agricultural production systems had an exciting impact on agriculture and brought visions of people being forever freed from hunger and drudgery.
8) In the 1950's farm management and integrated production systems came into importance. Pest management, plant breeding and improved varieties, fertilizers and improved soil management, water management and soil and water conservation, understanding of crop physiology and climate variables, and farm mechanization all provided their inputs into systems for optimum combinations for maximum food production.
9) Also in the 1950's the United States made major decisions and evolved programs for carrying its research and educational activities to help the developing countries increase food production. This was climaxed in the 1970's by development of world cooperation through the Consultative Group and its major international research centers. We have not yet, however, mastered the effective transfer of technology and its adaptation into the many national programs and, in turn, to its use on small farms of developing countries.

Agronomists can look with pride to advances over our nation's 200 years of gains in food production. The average yields of corn in the United States have increased four fold. The yields of wheat have tripled. Agronomic research has essentially made the soybean an important crop and a major source of oil and protein. Increases for other legumes and forage crops have been less spectacular, but technologies for these crops assures substantial yield gains in the future.

Europe and other developed regions of the world have had comparable gains in agricultural productivity. The benefits of improved agronomic practices have been extended to many developing countries. But even in our most optimistic dreams we can not assume the world battle for food supply has been won. With ever rising populations and the increasing expectations of earth's poorer peoples, the gains in food production have provided only modest improvements in satisfying human needs.

We can only assume that the substantial contributions of agronomists and other agricultural scientists to world resources will provide another few decades for mankind to learn to control population and to manage social problems. The spectre of hunger seems destined to remain with us. We have had our brief periods of optimism and despair as yields have increased and drought and energy limitations have wiped out food reserves. But malnutrition and the threat of famine remain close by. I am reminded of the rhyme:

As I was going up the stair
I met a man who wasn't there.
He wasn't there again today
I wish, I wish he'd go away.

Yes, we wish the spectre of hunger would go away. Agronomists have done much and can do more. But, the final solution of the world food problem will require full cooperation of all society in population control and an international commitment for the considerate and loving regard for the careful preservation and use of the world's resources and for the future of humanity itself.

LITERATURE CITED

Baeumer, K., and W. A. P. Bakermans. 1973. Zero tillage. Adv. Agron. 25:77–123.

Brown, A. W. A., T. C. Byerly, M. Gibbs, and A. San Pietro (ed.). 1976. Crop productivity—Research imperatives. Mich. Agric. Exp. Stn., East Lansing, Mich. and Charles F. Kettering Found., Yellow Springs, Ohio.

Carrier, Lyman. 1923. The beginnings of agriculture in America. McGraw Hill, New York.

Ennis, W. B., and W. D. McClellan. 1964. Chemicals in crop production. p. 106–112. *In* USDA Yearbook, Farmer's World. U. S. Government Printing Office, Washington, D. C.

Haise, H. R., and R. M. Hagan. 1967. Soil, plant and evaporative measurements as criteria for scheduling irrigation. *In* R. M. Hagan, H. R. Haise, and T. W. Edminster (ed.) Irrigation of agricultural lands. Agronomy 11:577–604. Am. Soc. Agron., Madison, Wis.

Hays, H. K. 1951. A half century of crop breeding research. Agron. J. 49:626–631.

Hilgard, E. W. 1906. Soils. Macmillan, New York.

Kelley, W. P. 1940. Cation exchange in soils. Am. Chem. Soc. Monogr. 109. Reinhold Publ. Co., New York.

McCalla, T. M., and J. T. Army. 1961. Stubble mulch farming. Adv. Agron. 13:126–196.

Promsberger, W. J. 1976. A history of progress in mechanization. Paper No. 76-1046. Am. Soc. Agric. Eng. Meeting, Lincoln, Neb. 27–30 June 1976, Mimeo.

Russell, E. W. 1973. Soil conditions and plant growth. 10th ed. Longmans, New York.

Stakman, E. C., Richard Bradfield, and P. C. Mangelsdorf. 1967. Campaigns against hunger. Harvard Univ. Press, Cambridge, Mass.

Tisdale, S. L., and W. L. Nelson. 1975. Soil fertility and fertilizers. 3rd ed. Macmillan, New York.

Warburton, O. W. 1933. A quarter century of progress in the development of plant science. J. Am. Soc. Agron. 25:25–36.

SUPPLEMENTARY READING LIST

Coons, G. H., F. V. Owen, and Dewey Stewart. 1955. Improvement of the sugar beet in the United States. Adv. Agron. 7:90–141.

Johnson, V. A., chairman. 1976. Genetic improvement of seed proteins. Board on Agric. and Renewable Resources, Natl. Acad. Sci., Washington, D. C.

Krantz, B. A. 1945. Fertilizer corn for high yields. N. C. Agric. Exp. Stn. Bull. 366.

Miller, M. F., and H. H. Krusekopf. 1932. The influence of systems of cropping and methods of culture on surface runoff and soil erosion. Mo. Agric. Exp. Stn. Bull. 177.

Patterson, Fred L. (ed.). 1976. Agronomic research for food. Am. Soc. Agron. Spec. Pub. 26, Madison, Wis.

Richards, L. A., and G. H. Wadleigh. 1952. Water and plant growth. p. 73–251. *In* Byron T. Shaw (ed.) Soil conditions and plant growth. Academic Press, New York.

University of California Food Task Force. 1974. A hungry world: The challenge to agriculture. Div. of Agric. Sci., Univ. of California, Berkeley, Calif.

Wischmeier, W. H., and D. D. Smith. 1965. Predicting rainfall-erosion losses from cropland east of the Rocky Mountains. USDA Agric. Handb. 282.

Wortman, S., J. Mayer, N. S. Scrimshaw, and V. H. Young. 1976. Food and agriculture. Spec. Issue, Sci. Am. 235(3):31-205.

Agronomists and Food—Challenges

J. B. Kendrick, Jr.

It's a challenge to write about our challenges because in this context when we talk about challenges, we are really talking about the future. Ambrose Bierce defined the future as "that period of time in which our affairs prosper, our friends are true and our happiness is assured" (Levinson, 1967). But for a writer, on occasions like this, it is that period of time to be avoided. It is a subject not to be trusted—a subject full of surprises, booby traps, and opportunities to make wrongheaded statements to be regretted later.

As the good life expands and the tide of rising expectations advance, so too does the anxiety in many quarters that the world may exhaust the resources that sustain that good life in our future, that all of us and our way of life, ultimately, may not survive at all. The statistics are not comforting. By the year 2000, according to some estimates, food consumption will double or triple; water withdrawals also may increase two to three times; fuel consumption will be up three times, iron and steel more than two times, fertilizers four to five times, and lumber more than three times. Can we meet these projected needs and demands? No one really knows.

One of our challenges is the fact that world realities are changing faster than our comprehension of the changes and much more rapidly than the social and political processes and institutions designed to cope with them. A generation ago most of us had a very different feeling about the future than we do today. In the 1962 *Yearbook of Agriculture,* the then Secretary of Agriculture, Orville Freeman, referred to new goals and responsibilities because we had reached the ". . .point where it is possible to produce enough so that nobody in the world need lack food and clothing. . ." (Freeman, 1962). We are not quite that confident today.

That was not very long ago, but it seems like another era. Then, there was very little discussion of such topics as the population explosion, oil spills, an energy crisis, free speech movements, women's liberation, or soil samples from Mars and the Moon. Even 5 years ago hardly anyone, including our agricultural experts, anticipated that by 1974 nearly all of the U. S. devel-

---

J. B. Kendrick, Jr. is Vice President for Agricultural Sciences at the University of California-Berkeley.

oped farmland would be under cultivation, that we would be without our comfortable cushion of grain reserves, and that world food conferences would be discussing the dimensions of the famine facing not millions, but hundreds of millions of persons in the next 25 years.

All this is to underline the fact that the world is on the move, the pace of change is accelerating, and any projected version of the future on my part is likely to be less dramatic than the oncoming reality. For example, as far back in history as we can go, man has been expanding in range and numbers. But until recent generations, that expansion was comparatively slow. Now, we are all familiar with the grim arithmetic that tells us the population bomb is ticking at a rate that threatens sooner or later—and probably sooner—to confound our efforts to preserve environmental quality and eliminate human hunger and misery.

We are quickly moving from an era of abundance and surplus to one of scarcity, a move that entails a fundamental revision of some aspects of our economic and social thinking. We can no longer allow the use of essential resources without explicit concern for the shape of the future. It is difficult for us to visualize the world population and resource stresses from the vantage point of our abundantly stocked supermarkets, super highways, luxury cars, and many leisure hours. Let me bring today's world into sharper focus by looking at it as a microcosm. I do not know whom to credit for this dramatic description but I want to share it with you because it so vividly portrays our relationship to the rest of the world.

> If all the world were reduced to a village of 1,000 people—in this village would be 60 Americans. The remainder of the world would be represented by 940 other persons. The 60 Americans would have half the income of the entire town. The 940 others would share the remainder of the town's income. Three hundred and thirty people in the town would be classified as Christians. Six hundred and seventy would not. At least 80 townspeople would be practicing Communists—that's more than in our whole nation. Seventy others would be under Communist domination. White people would total 303, while non-white would be nearly 700. The 60 Americans would have a life expectancy of 70 years. The 940 others could expect to live not more than 40 years. The 60 Americans would have as many possessions as the average of all the rest of the people in the village. The 60 Americans would produce 16% of the town's total food supply. Although Americans eat 62% above the maximum daily food requirement, they would either eat most of what they grow or store it for the future at enormous cost.
>
> Since most of the 940 non-Americans would be hungry most of the time, it could lead to some ill feelings toward the 60 Americans, who would appear to be enormously rich and fed to the point of sheer disbelief by a great majority of other townsfolk.
>
> Of the 940 non-Americans, 300 would have malaria, 85 would have Shisto Somiasis, 3 would have leprosy, and 45 would die this year from malaria, cholera, typhus, and other infectious diseases. One hun-

dred and fifty-six will die from starvation and malnutrition. None of the 60 Americans will ever get these diseases and will probably never worry about them.

The 60 Americans would each spend $87 a year on liquor and tobacco, but less than $20 for the drugs needed for the finest medical care in the world, and they would loudly proclaim that medicine costs too much. We are a very interesting people!!!

Yet, the world needs food, and although we can produce food cheaper than anyone else, have more than we need, and spend less of our income for food, there are many Americans opposed to selling our food to anyone. How long are the 940 other people going to tolerate the 60 Americans?"[1]

That is a picture of the world today. As population increases phenomenally in Asia, Africa, India, and South America, it is a sobering thought to visualize the place Americans will have in that village of 1,000 people in the year 2000—only 24 years hence.

The central challenge of agricultural scientists is to find ways to produce unprecedented amounts of food by the year 2000 (one-third more than we produce now), to feed 100 million more people in the United States alone, and to meet demands of the world population now increasing by 80 million people per year. If the problem of balancing population and food supply is not solved, the challenge of producing an adequate food supply may transcend all others.

We have the capabilities and the potential to meet that challenge, but if we really expect to pull it off we will have to deal with some other challenges, and we will have to begin dealing with them right now. There are still more unsolved physical, biological, chemical, and engineering problems of the kind that agricultural scientists solved in the past to bring about the impressive increases in food production already achieved. But now there are also new problems that have come with the changing world realities I mentioned earlier—issues that involve intangibles, issues that stem from attitudes and values, and ideologies that are as difficult and as important to solve as the physical, tangible problems.

Few of society's investments have provided a payoff to equal the dividends returned from agricultural research and extension. The remarkable revolution in agriculture, from subsistence to abundance is very largely the handiwork of the agricultural sciences. According to one study, research and extension can be credited with 60 to 70% of agriculture's increased efficiency and 80% of its growth for the period 1929 through 1972 (Pavelis, 1973).

Growing affluence and population make it imperative that increases in agriculture's efficiency and performance continue as rapidly in the period ahead. If we can do that it will be accomplished with the tools we have used in the past—research and education. The list of problems for science to solve is not endless but it is long: how to turn human and animal wastes into re-

---

[1]Anonymous.

sources, how to increase yield potentials of crops and animals, how to reduce crop and animal losses due to diseases and pests, how to prevent post-harvest loss of crops caused by spoilage, how to process and distribute food more efficiently, and how to do all these things without depleting the world's land, water, and energy resources.

While increased productivity alone is not the total answer to the world's food supply problem, it along with population control is one of the central and essential elements. Large and sustained increases will be needed if we are to be able to feed a population of 7 billion people by the year 2000. But it is clear that increases of that order cannot be achieved with present knowledge and technology. The yields from existing technology and conventional techniques are beginning to level off. The lavish application of fertilizers, chemicals, and energy is no longer the sure and easy way to boost yield per acre. And even if the effectiveness of these ingredients had not begun to level off, their availability in the future is by no means certain. Our finite fossil fuel supplies are not only subject to depletion but to the uncertainties of international politics. And certainly they have been and will in the future be subject to dramatic price increases. The energy component of agriculture may well become a significant limiting factor. Modern agriculture depends on energy to run the machines, pump the water, process the products, transport them to market, and provide the base for fertilizers and agricultural chemicals. Shortages and high prices of fuel could seriously curtail our production, and could spell disaster to the poorer countries. India's 1974 fertilizer shortage, for example, was credited with costing that country 9 million metric tons (10 million tons) of badly needed wheat.

As we approach the time when existing knowledge and technology are "used up," the resources and materials needed for production are no longer abundant, and all easily available arable land is under cultivation, it is obvious that we must find new approaches to increase agricultural productivity. Indeed, if projected future needs are accurate, we will not only need new approaches but *revolutionary* approaches—a few breakthrough-type discoveries comparable to the introduction of chemical fertilizers or the development of artificial insemination—in addition to advances built on the application of many small gains in technology. Our best hope for achieving this kind of increase in productivity lies in harnessing the untapped biological potential of plants and animals. I think you are all familiar with the broad general areas considered the most promising for expanding research efforts.

The most obvious one, of course, and perhaps the one with the greatest potential payoff, is research designed to improve photosynthetic efficiency. Agriculture in all its forms worldwide is the greatest single source of converted solar energy and is the only industry which produces more energy than it consumes. The other primary source, the energy stored in the earth in the form of coal, petroleum, and uranium, is limited and diminishing. This vital biochemical process deserves our attention because it is the world's most important energy producing process, and it is a *renewable* source. Even our most efficient crops convert only a tiny percent of the sun's energy into the chemical energy used to produce our food and fiber. We need to learn how to increase the energy conversion efficiency of this process because the impact on our food supply would be prodigious. Nor would the impact be confined solely to our food supply.

Another research area of great promise is the development of bacterial strains capable of fixing nitrogen in the root environment of our major food plants. This kind of biological nitrogen fixation could drastically change and reduce the use of increasingly expensive fertilizer and save the enormous quantities of petrochemical stock needed to produce it, enhance plant growth, and minimize the release of nitrogen into streams and groundwater. According to one estimate, if only one-fourth of the legume's nitrogen-fixing capacity could be transferred to the major grain crops, grain yields might be increased by 50% (Brown, 1975).

We all have our favorite lines of basic research needed to bring about fundamental botanical transformations and dramatic upward curves in crop yields. Long lists of other targets could be developed for major inputs of basic science to:

1) open the possibilities for creating genetic or other biological improvements in plants by applying cell, tissue culture and DNA recombination techniques;
2) focus on the dozen or more plants that are basic sources of food for man to increase protein content, accelerate and control maturation, reduce dependence on fertilizer and water, improve adaptability to specific environmental conditions and broader climatic ranges, increase resistance to diseases and insects, and increase yields;
3) increase the efficiency of chemical fertilizers;
4) develop more efficient means of water utilization; and
5) develop new weapons to reduce the 30% food loss due to insects, weeds, viruses, fungi, bacteria, nematodes, and vertebrate pests each year.

There are important challenges to be met in adapting and expanding conventional technological research such as:

1) development of technologies for reducing energy use in producing and processing farm commodities,
2) development of multiple cropping systems,
3) extending soil mapping, classification, and information programs, and
4) gathering, evaluation, and preservation of germ plasm stocks.

The preceding are examples of challenges to be met as we strive to maximize agricultural productivity for the future. There are equally urgent problems to be *minimized* if we are to enjoy that future.

The waste disposal problem, for example, increases in magnitude along with increases in production and consumption. As our wastes exceed nature's capacity to assimilate them, we face prospects of irreversible environmental damage caused by contaminated water, air, and soil. The increasingly difficult task of disposing of our enormous quantities of waste materials and the growing number of regulatory constraints complicate our efforts to increase agricultural production and food processing.

Like other challenges we face, the waste management problem calls for inputs from many branches of science integrated in a systematic, large scale effort. With new technical processes and much more knowledge than we have now, the looming mountains of waste can be reduced and turned into assets. There is food, feed, fertilizer, fuel, and other products of value to be recovered for a society faced with diminishing resources, yet accustomed to an infinite supply of basic resources. Economic stresses cannot all be malevolent when they provide incentives to become more efficient and less wasteful.

Malthus thought that the resources available for agricultural production were land, water, and human and animal labor. He did not foresee the phenomenal transformations brought about by scientific and technical knowledge.

I have given you only a rather fast and sketchy survey of the *scientific* challenges for two reasons. First, because many other people who are closer to the field than I am can fill out the substance and the specifics better than I can. Second, unless and until we meet some other challenges outside the agricultural sciences, any expanded search for basic knowledge–which should have long since been launched–may be deferred indefinitely. That search may be deferred until the realities of the food situation move to the crisis stage.

I'd like to look at some of the nonscience challenges because my combination of academic and administrative experiences gives me at least some qualifications in this area, and because I think they are of crucial importance to everyone of us. During the past nearly 9 years, in particular, I have had regular contact with the *users* of the results of research, as well as with the state and federal agencies that fund research, and I have had the opportunity to work on agricultural issues at the national level. I believe I have some feel for the economic, social, and institutional problems and issues which I want to consider now.

One problem–one challenge–is a change in public attitude, a change that will affect the progress and prospects of agronomy and other agricultural sciences in the years ahead. Unlike medicine or law, for example, agronomy and related sciences are primarily publicly-supported professions and, let us face it, society is no longer willing to accord unquestioned number one priority to education and research.

Increasingly, the public is asking pointed questions regarding the relevance of higher education and research, and the barometers I read do not indicate any shift or improvement in this new climate any time soon. It is not a friendly climate for "elitist" institutions like universities or for the researcher who would pursue knowledge for its own sake.

The change in climate can be traced, in part, to a fairly general disenchantment with sciences, and for some, a desire to retreat from technological civilization entirely, brought on by undesirable side effects of technology.

And as the scale and complexity of the scientific enterprise grows, there is a corresponding decrease in public understanding and appreciation of that enterprise. Another factor is the growing competition for the tax dollar, the upward spiraling costs of providing human health and welfare benefits, educational and economic opportunities, and economic security are now having an effect on the economic stability of some of our large cities, and on governments at all levels.

State and federal appropriating bodies are asking two sorts of questions: (i) What is the payoff from the projects they support and who benefits from them? and (ii) How do we rank them among the issues and priorities that include the environment, the inner city, race, poverty, the consumer, energy, education, and the general quality of life? In an increasingly urban society, these questions will become increasingly insistent for the publicly supported agricultural sciences. To the public, these problems are real, visible, and urgent; to the urban majority, the problems and functions of agricultural research are remote and seemingly irrelevant. Within the university we can expect the same demands for accountability and relevance and expect to have our programs evaluated in the context of the university's budget against other programs addressing the issues of affirmative action, equal opportunity, the urgent problems of the urban life, and the rising cost of everything. Deans, directors and vice presidents must be able to justify all projects with convincing justifications and clearly identified benefits if they are to sustain their programs.

In California, budgets for university agricultural research have not been programmatically augmented in ten years. The recent increases in dollar amounts have not even kept pace with the increasing complexity and rising costs of doing research. The federal establishment's record of investment in science has been equally dismal. Since 1967, appropriations for research have not kept up with inflation. These facts of life are indicators of a fundamental change in attitudes of a society that has become increasingly complex and urbanized. Our society will not become less urban or less complex and the prevailing climate of attitudes I have described will not be changed easily, or soon.

One of the most unfortunate aspects of this money crunch is its effect on fundamental research. Times of austerity and budget stringencies reinforce the governmental tendency to favor the *immediate* problem and the program with the quick payoff and to postpone long-term needs. We need to tackle right *now* the kinds of problems that will yield results 5, 10, 15 years from now. Research to produce the kind of basic knowledge we will need to undergird our future food production technology requires lead time, carefully planned, large-scale multidisciplinary programs.

According to the report on *Enhancement of Food Production in the United States* published by the National Academy of Sciences, "The USDA–SAES complex has not adequately funded basic research relating to biological processes that control crop and livestock productivity." A formerly substantial basic research effort has "virtually disappeared," and "consequently the agricultural research system now faces inadequacies in fundamental knowledge about photosynthesis, nitrogen fixation, crop and livestock protection, water and nutrient efficiencies, genetics, biochemical aspects of handling and processing of crops, livestock and fish, and plant-soil-water relationship"—a rather damaging indictment (Wittwer, 1975).

Over the years, the Congress has done a great deal to advance the development of scientific research, but the declining appropriations for basic research in the past decade has had, and will have, serious consequences. We need an enlightened policy toward basic research to the end that we can anticipate and prepare for future problems before they reach the crisis stage and political pressures force the construction of wasteful, jerry-built crash programs to cope with them.

These past years, anyone who attended or read reports of the congressional hearings concerning agricultural research might well begin to have an "Alice-In-Wonderland" sense of unreality. Informed, forceful, and seemingly convincing testimony was presented at great length proving that increasing food production is a central bulwark for our economy, necessary to prevent massive world malnutrition, and preserve world peace, and that agricultural research is the key to that kind of production. All the expert witnesses (representing university agricultural scientists, the agricultural industry, and the USDA), are united in pointing out that large increases in research funding are necessary and urgent. Although these events are called hearings, somehow, not enough people hear our message and nothing significant has happened yet to our funding.

There is a touch of paranoia in the agricultural science community, after all, 40% of the federal research and development funds went for agricultural research in 1940, and in 1970 we were down to less than 2%, but the current unfriendly climate seems to have affected other segments of the nation's research establishment as well.

A report by the National Science Foundation's National Science Board presents some indications that all is not well with American science. The Board's 1975 annual report stated the following.

1) During the last decade the proportion of the U. S. Gross National Product spent for research and development declined steadily, and during the same period there was substantial growth in West Germany, Russia, and Japan.
2) The number of scientists and engineers engaged in research and development declined from 558,000 in 1969 to 528,000 in 1974. In most major countries the trend was in the other direction.
3) A significant increase in the number of U. S. patents issued to foreigners (30% of the total), indicated that the number of patentable ideas is growing faster in other countries than in the United States (Hackerman, 1975).

Derek C. Bok, president of Harvard University, also spoke of dangers to the research university posed by this new and unfavorable climate. In an editorial in *Science,* he deplores the threat to experimental science due to shortage of funds, and the neglect of long-range international needs in such fields as basic biological research devoted to agriculture brought on by pressures to use research funds for urgent domestic problems, such as cancer and energy. He suggests that the career opportunities of younger, promising university scientists are being limited by financial stringency and the work of established investigators hampered by instability in federal funding and by the increasing amount of red tape involved (Bok, 1976).

It is clear that science and government are closely and increasingly interrelated, and it is also clear that the relationship needs improvement for the good of both parties. Some strong words by another university president, Harold L. Enarson of Ohio State University, speaking to Ohio's congressional group, seems particularly relevant in this connection. I quote: "A fundamental change is taking place in the relationship between Washington and the nation's colleges and universities, a change I find deeply disturbing. Once we were partners working together to solve national problems. Now we view each other with suspicion, almost as adversaries. We overregulate on one hand and overreact on the other. We have placed our partnership in peril. And if it is to be restored, it urgently needs our attention and understanding."

It is an interesting academic exercise to describe the scientific challenges and promising programs with which we hope to usher in a new agricultural revolution, but, as I said earlier, it might be more constructive, and I think it is more urgent, to direct our attention first to the state of health of the publicly-supported research enterprise. The catalogue of symptoms previously described makes clear that its condition is not all that it might be—not robust enough for the tasks ahead.

I have no sure cure to offer, but I do have a deep conviction that, for the good of the human enterprise and for its own self-interest, the community of science had better get its act together. Finding the means to improve our communication with the public is one challenge. All science must ultimately serve society, for it is society that provides its reason for being and most of its support. But science can serve the public well only if it *is* supported and independent, and this condition will prevail only *if* there is an informed public.

We cannot all become experts in the science of politics, but we had better become more expert than we are in the politics of science. The scientists' understanding of government must be improved along with the public's understanding of science. As agricultural scientists we have some responsibility to prevent, if we can, the neglect of *our* needs in favor of more politically immediate demands. We have more information and a better understanding of what it will take in terms of resource use, manipulation of the environment, and sheer scientific and technological achievement, to produce a food supply of the magnitude needed by the end of this century. Somehow, someway, we must find the means to increase our impact at the decision-making level.

The first step, it seems to me, is to recognize that we are in a new era, in a new league with different values, a less favorable environment, and harsher competition for the tax dollar. Science and technology are the foundation for our prosperity, our health and our food supply, but we do not have a real science policy or a consistent, rational provision for the science program. The science community must face the reality that our science program is in jeopardy for lack of public understanding and support.

The national community of scientists is a formidable aggregation of intellect and influence, but its power and its strength is divided, compartmentalized and diffused. We need a stronger voice. We need to think of ways to organize, focus and bring to bear, not only the collective voice of scientists and their professional organizations, their universities and university organization and associations, but also the various allies outside the universities who share an interest in the quality and advancement of sciences. The recent restoration of the office of Science Advisor to the White House does show promise of improving science's role in public affairs, but there still is no evidence that food and fiber research are high on the list of national priorities.

We must recognize our interdependence, seek alliances, involve representatives of our various publics in advisory and participatory roles, and join forces with other interest groups to broaden our base. We must also recognize our strength and our values. Science brought about the revolution in agriculture that made possible our present abundance, and it is science that must be depended upon to bring about the next one. The community of science can no longer be reticent; it must forcefully convey the message that society is also dependent on us.

Given the scope and the means, science could very well make Ambrose Bierce's somewhat sardonic definition of the future come closer to the mark than the grim forecasts of present day handwringers and prophets of doom. If 90% of all that science has learned has been learned during our lifetime, think what we can learn and what doors we can open in the years ahead. After all, unlike the dissipation of matter and the depletion of resources, scientific creativity grows from within and increases man's resources and expands his universe.

LITERATURE CITED

Bok, D. C. 1976. Universities and national research policy. Science 193(4257):955.

Brown, H. S., chmn. 1975. Interim report: World food and nutrition study. World Food & Nutrition Study Comm., Natl. Acad. of Sci., Washington, D. C.

Freeman, O. L. 1962. Forward. p. v. *In* The yearbook of agriculture, 1962. USDA, 87th Congress, 2nd Session, House Document No. 279, Washington, D. C.

Hackerman, Norman, chmn. 1975. Science indicators 1974. 7th Ann. Rep., Natl. Sci. Board, Natl. Sci. Found., Washington, D. C.

Levinson, L. L. 1967. Webster's unafraid dictionary. McMillan Co., New York. p. 94.

Pavelis, G. A. 1973. Energy, national resources and research in agriculture. Economic Research Service, USDA, U. S. Government Printing Office, Washington, D. C.

Wittwer, S. H., chmn. 1975. Enhancement of food production for the United States. Board on Agric. & Renewable Resources, Natl. Acad. of Sci., Washington, D. C.

The Role of Wheat in America's Future

V. A. Johnson

I would like to share with you some of my thoughts about the role of wheat (*Triticum aestivum* L.) in our future. This requires examination of the world food and people situation, because wheat is the world's major producer of calories and protein.

More wheat is produced annually than any other food or feed crop. Annual world production of 360 million metric tons exceeds that of rice (*Oryza sativa* L.) by 40 million metric tons (Harlan, 1976). Maize (*Zea mays* L.) and potatoes (*Solanum tuberosum* L.) provide the next largest tonnages. An estimated two-thirds of the world's 4 billion people rely on wheat and rice as dietary mainstays.

The United States is the world's largest and most reliable exporter of food and feed grains. Together, the United States and Canada have a net grain trade of 96 million metric tons annually (Wortman, 1976). This compares with only 10 million metric tons for Australia and New Zealand, the world's next largest grain exporters. The remainder of the world's countries with exception of Argentina and Republic of South Africa encounter annual deficits and frequently must import grain. Wheat is the major export cereal from North America, Australia, and New Zealand. Clearly these countries compose the wheat granary of the world.

The United States produces only 12.7% of the world's wheat and yet it is the largest international supplier of wheat. An obvious explanation is the continuing favorable ratio of population to wheat production in the United States. The same explanation adequately accounts for the grain export status of Canada, Australia, and New Zealand.

An exponential population increase shown in Fig. 1 is a grim reminder of our precarious future. The world's population already has passed the 4 billion mark and, in another 10 years, will have reached 5 billion people. Obviously, there can be no lasting solution to the food problem without a "leveling off" of population. Compared with the 1 billion population increase projected for the next decade, the size of the North American granary is small indeed. At its current level of production, American agriculture can feed only an estimated 50 million more people than now live in the United

---

V. A. Johnson is a research agronomist for the USDA/ARS at Lincoln, Neb.

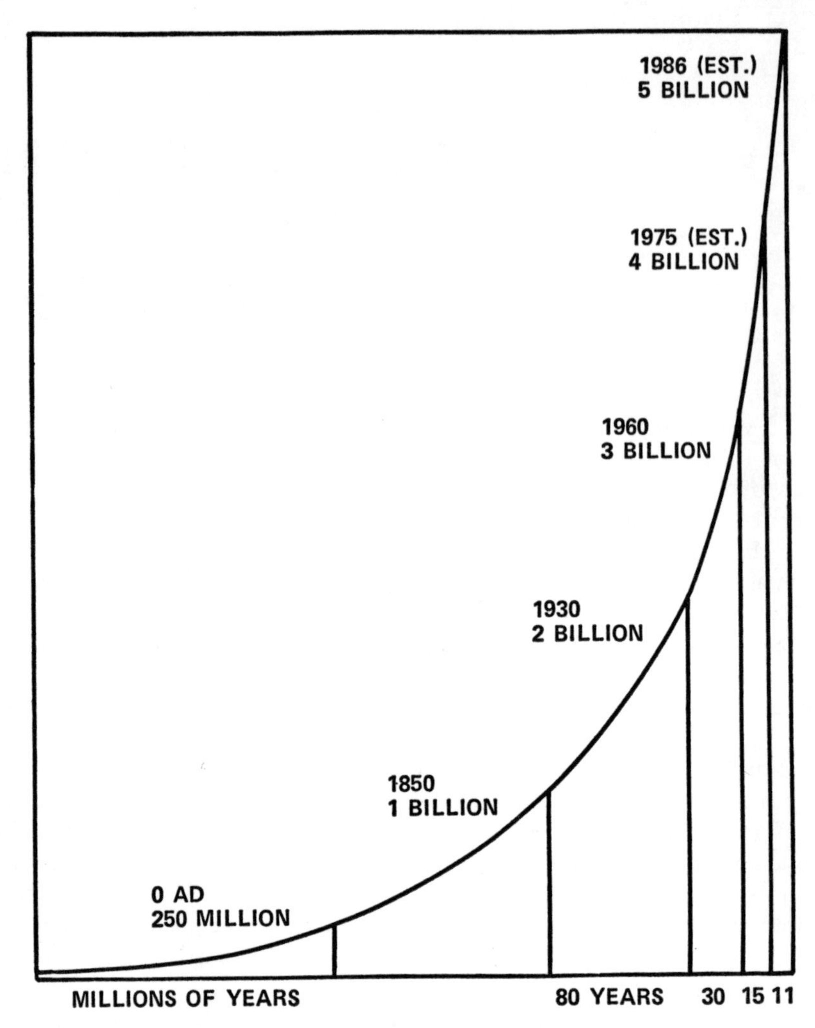

Fig. 1. The world's population is increasing exponentially. By 1986 it will have reached 5 billion people (Source: United Nations. Taken from *The growing challenge—protein cereal products for world needs.* A. D. M. Milling Company, 1972.)

States at the nutritional level we enjoy in this country. An encouraging note in this otherwise grim outlook is recent evidence that world population growth may be somewhat less than projected due in part to smaller families in the United States and the People's Republic of China which together possess one-fourth of the world's people.

World cereal production since 1960 has steadily increased. But consumption of cereals has increased even more with the result that there is a growing cereal deficit. Even larger cereal deficits are projected for the next decade. Confronted with such a projection, one can reflect on Malthus' 1798 prediction that "population must eventually overtake food production capability of the world" (Wortman, 1976). Was Malthus really wrong, or is it only the "timetable" that has been in error? Which ever the case, there can be little doubt that new agricultural breakthroughs together with effective reduction of population growth will be necessary to avert eventual mass starvation in many parts of the world. The United States as the major exporter of wheat must play a key role in this effort.

How will increased production of wheat be accomplished in the United States? Will it come through expanded acreage? This seems unlikely since it would require diversion of cultivated acres from other crops to wheat. Without major increases in the price of wheat relative to other crops, wheat cannot compete with these crops in the more favorable production areas. Only in the plains of the central U. S.–particularly the high plains where there are few, if any, adequate substitutes for wheat–and in the Pacific Northwest can wheat be expected to continue domination of the production area.

If this assessment of future U. S. wheat acreage is correct, then the needed production increases must come from higher yields. Can we anticipate continued advances in the yield of wheat following established conventional breeding procedures, or must we resort, as some believe, to new unconventional breeding techniques?

In Fig. 2 the mean regional performance of experimental varieties of hard winter wheat grown in the Southern Regional Performance Nursery[1] has been plotted against that of the 'Kharkof' check variety for the 13-year period since 1963 to show breeding progress under the largely rainfed production environment of the southern and central plains. Additionally, the yield of the most productive variety or experimental line each year based on its regional average is shown. Steady general improvement of yield is indicated by the upward trend of the nursery mean yield of all experimental varieties.

During the 5-year period from 1963 to 1967 the 'Scout' variety and 'Selections' from Scout dominated nursery yields. A Scout yield threshold of approximately 130% of Kharkof is indicated. In 4 of the next 5 years from 1968 to 1972 a new variety 'Centurk' was the highest yielder. A Centurk yield threshold of 140% of Kharkof and 10% above Scout is suggested. Since 1972, you will note the apparent emergence of a group of highly productive wheats that suggests a possible new yield plateau of more than 150% of Kharkof. The wheats in this group are derived from crosses of semi-dwarf wheats (mostly high yielding spring types from Mexico) with adapted hard winter varieties. However, there are some trade-offs. These new wheats appear to lack winterhardiness to permit their safe production north of Kansas and Colorado and all are inherently lower in protein than currently-grown varieties. Whether the high yield potential of the wheats can be retained in subsequent crosses with winter wheats to recover needed winterhardiness remains a question.

---

[1]Replicated winter wheat nursery grown at 27 sites in 11 states in which advanced experimental lines are evaluated prior to their release. Data from USDA, ARS Reports, 1963–1976.

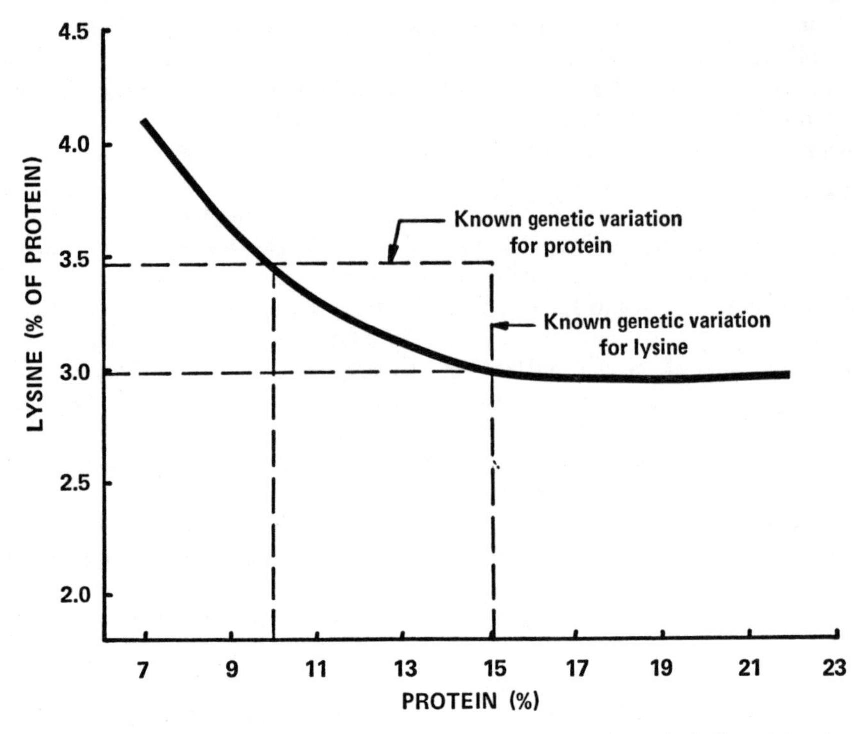

Fig. 2. Increasing productivity of hard red winter wheat varieties is indicated by the mean yield of experimental varieties tested in the Southern Regional Performance Nursery since 1963. A yield threshold of 155% of Kharkof may have been reached since 1972.

Most important, in my opinion, the high yields of these new wheats strongly indicate the existence of useable genetic variation for yield under the variable and hazardous rainfed production environment of the southern and central U. S. plains wheat area. They offer a challenge to wheat breeders to devise innovative new approaches to identify and capture this variability in high-yielding new wheat varieties. Also, these new winter wheats lend strong support to the massive systematic program of winter by spring wheat hybridization currently in progress at the International Maize and Wheat Improvement Center (CIMMYT) and Oregon State University as a viable approach to further significant improvement of both winter and spring wheats.

Continued genetic improvement of the yield potential of wheat varieties must be accompanied by production practices that permit maximum expression of genetic potential under the sub-optimal rainfed production environments of the United States and other countries in which most wheat is grown. In fact, full application of the "total production package"[2] concept to such production environments may yield the largest returns of all. The

[2] Refers to a complete management system in which genetic improvement of varieties is combined with application of full management practices to achieve maximum yields.

"Green Revolution" is a beautiful example of the application of the total production package to achieve significant advances of wheat yields under irrigation and areas of adequate rainfall. The production package concept, in my opinion, has not yet been adequately applied to the sub-optimal rainfed wheat areas of the United States and other countries. The package, to achieve its maximum usefulness, necessitates attention to all constraints to maximum expression of the yield potential of the varieties grown, i.e., water retention and conservation, weed control, improved wheat management practices, soil fertility, and control of diseases and insects, some of which have received little attention in the past. All of these must be achieved within the limits of potentially the most serious production constraint of all—that of energy or lack of energy.

Will hybrid wheat play an important role in the future? Early expectations for hybrid wheat have not yet been realized. The cytoplasmic sterility-genetic restoration system used thus far by breeders is clumsy and laborious. Some breeders believe that it interferes with yield. The number of hybrid combinations evaluated to date is relatively small and none, to my knowledge, has exhibited the level of heterosis for yield required to be economically competitive with the best wheat varieties. However, to conclude, on the basis of the number of hybrids tested, that there is insufficient heterosis for yield in wheat, is not warranted.

Two developments with implications for hybrid wheat deserve mention. The recent identification of chemicals with apparently reliable male gametocidal properties opens the way for development and evaluation of large numbers of new hybrid combinations from which an answer to the question of "useful" heterosis in wheat, should finally be provided.[3] Further, such chemicals, if proven to be adequate for commercial hybrid production, should cut through the laborious and time-consuming task of male sterile conversion and manipulation of genes for fertility restoration. A second significant development is the induction and isolation of a recessive gene for male sterility from wheat (*Triticum aestivum* L.) nuclei substituted into *Aegilops squarrosa* cytoplasm by researchers at North Dakota State University (Frankowiak et al., 1976). Theoretically, this could provide a much-simplified system for male sterile development and fertility restoration.

I believe that improvement of the nutritional quality of wheat must become an important component of future wheat improvement in this country and elsewhere. Substantial genetic variation for level of protein in wheat grain has been identified in the last two decades and many breeding programs now are utilizing these genes. Agricultural Research Service (ARS)/Nebraska research has produced evidence that useable genetic variation for protein may amount to as much as 5 percentage points (Johnson et al., 1975). Unlike the diploid cereals corn (*Zea mays* L.), barley (*Hordeum vulgare* L.), and sorghum (*Sorghum bicolor* L., Moench), genes in wheat with large effect on lysine have not been identified.

---

[3]Verbal communication with J. E. Stroike, Rohm and Haas Company, Research Laboratories, Spring House, Pa.

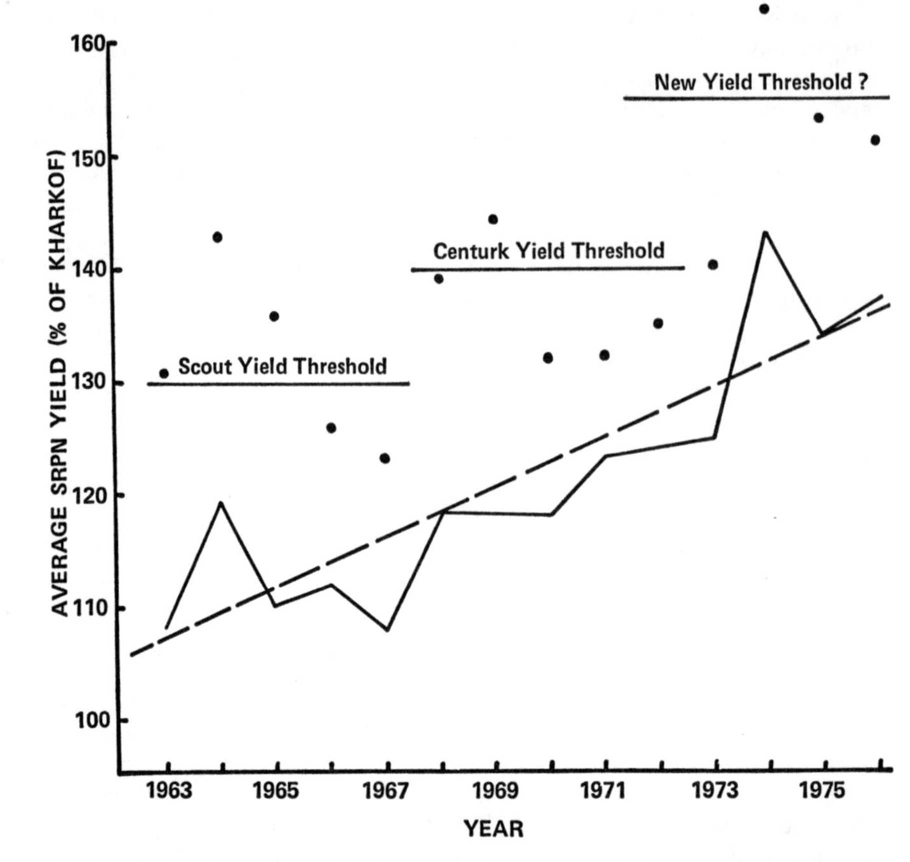

Fig. 3. Lysine expressed as % of protein decreases as protein content of wheat is increased. Known genetic variation for lysine of 0.5 percentage point combined with known genetic variation for protein of 5 percentage points will permit development of high protein varieties with as much lysine as lower protein varieties.

We have identified genetic variation for lysine in wheat of sufficient magnitude to overcome the normal depression of lysine associated with increases in protein content (Fig. 3) (Johnson, 1977). The broken vertical line extending above the regression line of lysine on protein indicates the amount of known genetic variation for lysine with which we are working. Utilization of this variation in combination with known genetic variation for protein shown by the broken horizontal line above the regression line, will permit development of wheat varieties 5 percentage points higher in protein content, and with lysine comparable to that of lower protein wheats. Productive experimental lines that exhibit this combination of protein and lysine already have been developed in the program at Nebraska.

Some agencies are employing mutation induction to increase genetic variability for lysine in wheat. In my opinion the polyploid genetic constitution of the economic species of wheat (*Triticum durum* and *Triticum aestivum*) imposes a strong constraint to further improvement of quality of wheat protein by this means.

Examination of the future of wheat in the United States would be incomplete without consideration of interspecific and intergeneric hybridization. Translocations and whole chromosome substitutions have been employed in Missouri, Oklahoma, and South Dakota (Sears, 1956; Sebesta & Bellingham, 1963; Sebesta & Wood, 1977) to successfully transfer resistance to leaf rust, wheat streak mosaic, and greenbugs from related species and genera to hexaploid wheat. I believe that, in the future, there will be an acceleration of this technique for transfer of useful genes to wheat from related species and genera.

The hybridization of durum wheat and rye (*Secale cereale*) followed by chromosome doubling to achieve the stable new plant form "triticale" (X *Tritico secale* Wittmack) is a notable recent achievement in plant breeding. The future role of triticale in this and other countries is difficult to predict, but there is optimism, based on the performance of some triticale lines, that it can make a significant contribution to future world cereal production.

Experts agree that breakthroughs in wheat productivity will be necessary to keep pace with the increasing world need for wheat and other food grains. Much improved understanding of the physiological processes associated with wheat productivity and quality will be needed. New innovative approaches to the development of genetic variability must be undertaken. Techniques for differentiation of callus to produce complete plants, isolation and fusion of protoplasts, and direct manipulation of DNA may represent powerful future laboratory techniques for useful modifications of the wheat plant and must be adequately researched. It is important, however, that they be properly identified as potential tools in our total arsenal of techniques for wheat improvement. They can only complement conventional wheat breeding; they cannot substitute for it. Results of innovative new techniques such as these, however spectacular they might be, will not increase wheat productivity until the breeders and geneticists combine them with the many agronomic, morphologic, and quality characteristics required of modern wheat varieties.

The maxim that, "Tomorrow belongs to those who plan for it today," has never been more applicable to a situation than the one that now confronts the world's agronomists. The need for innovative research in plant improvement has never been greater; it requires action now—not tomorrow! The challenge to us as agronomists is unparalleled, but so are the opportunities. To be sure, there are many constraints, some of which are beyond our control. But I'm not as concerned with these as I am with those constraints that may be imposed by our own lack of vision, by our failure to perceive the true nature and seriousness of the challenge, and by our reluctance to break down barriers to communication and joint action with scientists of associated disciplines.

If those of us in wheat research truly accept this challenge, then the role of wheat in our country's future will be fully as significant as was its role in our heritage.

LITERATURE CITED

Franckowiak, J. D., S. S. Maan, and N. D. Williams. 1976. A proposal for hybrid wheat utilizing *Aegilops squarrosa* L. cytoplasm. Crop Sci. 16:725-728.

Harlan, J. R. 1976. The plants and animals that nourish man. Scientific American 235(3):89-97.

Johnson, V. A. 1977. Wheat protein. p. 371-385. *In* Proc. Int. Symposium on Genetic Control of Diversity in Plants. 1-7 Mar. 1976. Lahore, Pakistan. Plenum Publishing Corp., New York, N. Y.

Johnson, V. A., P. J. Mattern, and K. P. Vogel. 1975. Cultural, genetic, and other factors affecting quality of wheat. p. 127-140. *In* Bread. Applied Science Publishers, Ltd., London, England. 358 p.

Sears, E. R. 1956. The transfer of leaf rust resistance from *Aegilops umbellulata* to wheat. Brookhaven Symposium on Biology 9:1-22.

Sebesta, E. E., and R. C. Bellingham. 1963. Wheat viruses and their genetic control. Proc. 2nd Int. Wheat Gen. Symposium. Hereditas Suppl. Vol. 2:184-203.

Sebesta, E. E., and E. A. Wood, Jr. 1977. Origin of 'Amigo' (C.I. 17609) greenbug resistant wheat germ plasm. *In* V. A. Johnson (ed.) Proc. 14th Hard Red Winter Wheat Conference. 8-10 Feb. 1977. Lincoln, Neb. ARS unnumbered publication.

Wortman, S. 1976. Food and agriculture. Scientific American 235(3):31-39.

Wheat—Its Role in America's Heritage

John W. Schmidt

The role of wheat (*Triticum* sp.) in America closely parallels that of its immigrant people since it accompanied the European immigrants. Once in this country it was a favorite crop of the westward-moving pioneers. For them wheat was important as a food staple and as a source of money, or was used in lieu of money—even for wages.

The earliest generally accepted date for the first wheat seeding in the United States was in 1602 by Gosnold on Cuttybank, one of the Elizabeth Islands in Buzzards Bay, Mass. (Flint, 1973). However, Carrier (1923) reports that wheat was seeded as early as 1578 in Newfoundland, Canada by fishermen. The very early Virginia settlers tried wheat as did the Pilgrims whose first seeding failed. In New England wheat did not compete well with the native maize (*Zea mays* L.). Soil depletion, the introduction of the stem rust fungus, and the presence of a wheat midge led to the gradual disappearance of wheat from coastal New England and hastened its movement westward (Rasmussen, 1960, 1975).

In colonial America, Pennsylvania was the important wheat state (Carrier, 1923). There wheat replaced maize by the mid-1700's, and as late as 1849, Pennsylvania was still the leading wheat state (Flint, 1973). The first production census was taken in 1840; thus, records previous to that time are sketchy. According to the U. S. Department of Agriculture (USDA) (Reitz, 1954), Pennsylvania was the center of production in 1839; in 1849 and 1859 Ohio was the center of production; by 1869 it had moved to Indiana and by 1879 to Illinois. By 1889 it had moved to the Illinois-Iowa border, and spring wheat expansion moved the center of production to the Minnesota-Iowa border by 1899. The impact of hard red winter wheat (*Triticum aestivum* L.) production in the Great Plains moved the center of production to northeastern Nebraska in 1909. Since then, the center of production has been in northern Kansas with a slow movement westward through 1949, reflecting improved production especially in the

John W. Schmidt is an agronomist at the University of Nebraska, Lincoln, Neb.

Pacific Northwest. Currently, increased production in southwestern and western states tends to move the center of production somewhat in a southwesterly direction.

In 1840, the census reported that 2.3 million metric tons (85 million bushels) were produced (Flint, 1973). In the Centennial year of 1876 slightly more than 8.4 million metric tons (309 million bushels) were produced, of which 1.6 million metric tons (57 million bushels) were exported (Becker et al., 1941). In this Bicentennial year of 1976, 57 million metric tons (2.1 billion bushels) were produced (Statistical Reporting Service, 1976) and about 27 million metric tons (1 billion bushels) will be exported (Economic Research Service, 1976). Yield was 732 kg/ha (10.9 bushels/acre) in 1876 and 2042 kg/ha (30.4 bushels/acre) in 1976.

The figures above, important as they are, are not the interesting story of wheat in America. As America was the melting pot for people of many nations, so America was the melting pot for the world's wheat germplasm. In all of this, the people were important. Immigrants brought many kinds of wheats; others were introduced purposefully. Once here, wheat was nurtured, mixed, selected, and crossbred. Important cogs were the ordinary farmer, the farmer-seedsman-breeder, the researcher, and the politician. Edgar (1911) reports that George Washington wrote in his last letter that "as a farmer, wheat and flour are my principal concerns". The combined activity of these groups provided the basis for the unmatched varietal array of wheat types available today—a veritable cafeteria for the flour-milling and baking industry to choose from in providing the pasta, pastry, and bread for our tables.

In the confluence of germplasm materials that flowed to America during this Bicentennial period (our first 200 years), there were certain key or landmark introductions that stand out from the rest. Varietal names often mean little because synonyms abound, but they denote a recognizable type and are often tied to a geographical area of the United States. The discussion that follows will be by regions and wheat types, not chronological, and is based largely on the USDA wheat classification bulletins (Clark et al., 1922; Clark and Bayles, 1935, 1942, 1954).

NEW ENGLAND

There is little varietal information regarding the wheats grown in this area (Fig. 1). However, they were the earliest introductions by immigrants from western Europe and would be expected to be similar to the soft white Dutch, English, and possibly Swedish winter wheats of that period (Clark, 1936). Varietal identity of these wheats is obscure partly because of time, but also, because stem rust, the wheat midge, and later, Hessian fly, largely eliminated these varieties. However, the selection of "offtypes" in later-appearing varieties could trace to admixtures from these earlier wheats and thus they may be assumed to have some importance.

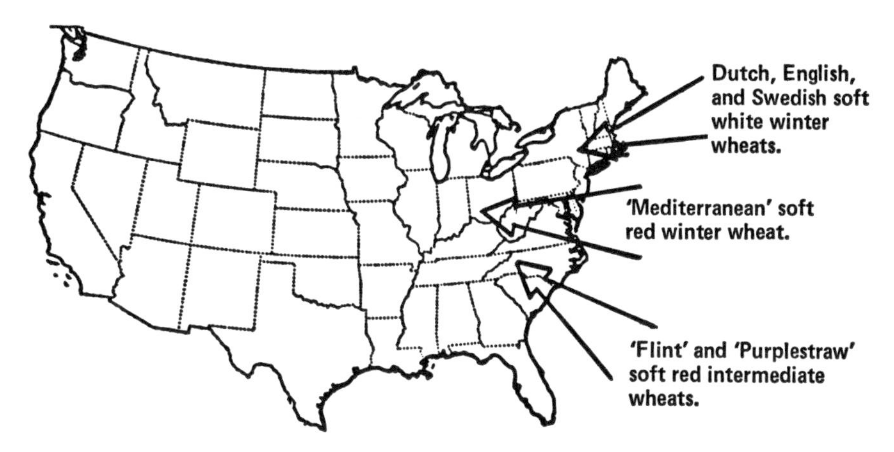

Fig. 1. U. S. Wheat Introductions.

SOUTHEASTERN STATES

Two very early introductions of undetermined origin provided the keys for wheat production in this region (Fig. 1). 'Purplestraw' provided the cornerstone for wheat production while 'Flint' ('Red May' or other synonyms, possibly tracing to old English Lammas varieties) provided important intermediate habit, soft red germplasm through such varieties as 'Redhart' to varieties in production today.

MIDDLE ATLANTIC AND CORNBELT STATES

In this region when you have said 'Mediterranean' you have about said it all (Fig. 1). The introduction of this soft red winter wheat variety is truly a landmark! Again, time and origin are uncertain. Well-known varieties stemming from selections of hybrids involving this land variety are 'Lancaster,' 'Fultz,' 'Fulcaster,' 'Trumbull,' 'Leap,' 'Leapland,' and on into the present Cornbelt varieties. Admixtures or natural hybrids within this germplasm may have also provided such an important variety as 'Goldcoin' in New York. Mediterranean filled the void created by the demise of other varieties due to susceptibility to the Hessian fly—probably introduced during the Revolutionary War.

SOUTHERN AND CENTRAL GREAT PLAINS

According to Quisenberry and Reitz (1974), the introduction of 'Turkey' wheat into the midlands in 1874 by the migration of the Mennonites from Russia, provided the cornerstone for the hard red winter wheat empire

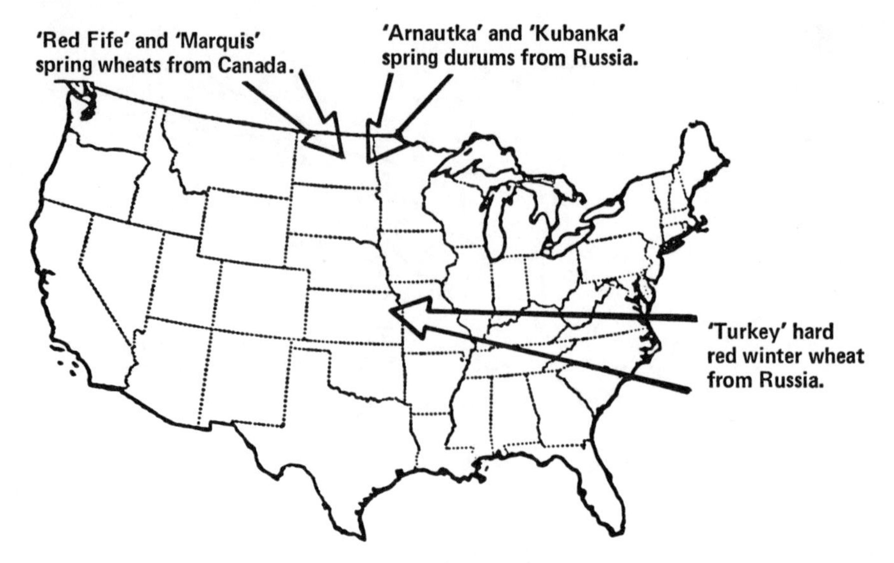

Fig. 2. U. S. Wheat Introductions.

that was to develop (Fig. 2). Although this land variety may have been introduced a few years earlier, the monumental effect this variety was to have on Great Plains agriculture originated with the 1874 introduction. Nearly 90 years later Briggle and Reitz (1963) reported the presence of Turkey wheat germplasm in the breeding of 45% of all common wheats and of 100% of hard red winter wheats grown in the United States in 1959. Turkey (and closely related 'Crimean' and 'Kharkof') wheat germplasm is today still the basis for most hard red winter wheat improvement programs. 'Cheyenne,' a selection from Crimean, provides the basis for the Nebraska wheat breeding program while 'Kanred,' another selection from Crimean, is present in numerous pedigrees. A hard red winter wheat released in 1971 was named 'Centurk' (Centennial Turkey) in commemoration of this important introduction.

NORTHERN GREAT PLAINS

1. Hard Red Spring Wheat

The introduction of 'Marquis' hard red spring wheat from Canada in 1912 by the USDA was to have an effect on the spring wheat breeding programs similar to that of the Turkey wheat introduction (Fig. 2). In contrast to the Turkey introduction, Marquis was not a pioneer spring wheat, for spring wheat was well established by 1912, when Marquis was introduced. However, its germplasm and production values were as important. Earlier, 'Red Fife,' one parent of Marquis wheat and also introduced from Canada, was an important variety in its own right. Fife, or Red Fife, originated from a single spring wheat plant selected from a winter wheat field which had been seeded near Glasgow, Scotland, with a sample of wheat supposedly of Russian origin.

2. Spring Durum Wheat

Two spring durum introductions are of importance (Fig. 2). One, 'Arnautka,' introduced by immigrants and the USDA, and the second, 'Kubanka,' introduced principally by M. A. Carleton of USDA, laid the basis for our spring durum industry. However, while both were important in showing that durums were adapted to the northern Great Plains, neither contributed conspicuously to the spring durum gene pool. That position is occupied by 'Mindum' durum, a single plant of durum selected from a Minnesota field of 'Hedgerow' common wheat!

PACIFIC COAST

1. Club Wheats

Production of club wheats in the Pacific coast states traces to introductions of club wheats along with other Spanish wheats from Mexico and Chile, probably by Spanish missionaries (Fig. 3). 'Little Club' was used extensively in crosses by W. J. Spillman to develop a number of hybrid clubs, notably 'Hybrid 128.' Hybrid 128 has provided important germplasm to present day club wheats. 'Big Club' provided important germplasm to the California wheat breeding program.

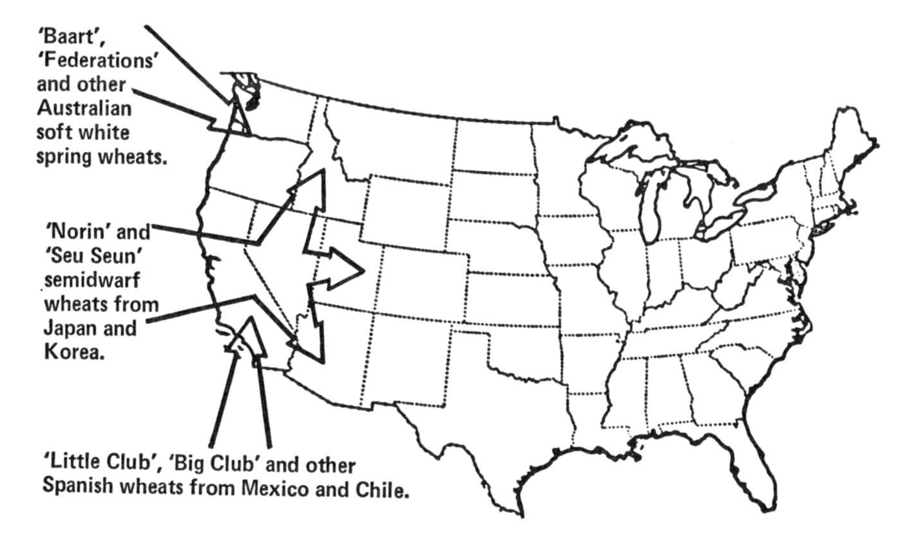

Fig. 3. U. S. Wheat Introductions.

2. Australian White Spring Wheats

White spring wheats such as 'Pacific Bluestem' were introduced from Australia into the Pacific Northwest in the mid-1800's (Fig. 3). While this and similar varieties set the stage for white wheat production in the Pacific Northwest, the introductions of 'Baart' in 1900 and the various 'Federation' varieties in about 1915, all from Australia, were more important. Federation especially has been an important germplasm contributor in the breeding of soft white spring and winter wheats in the region.

SEMIDWARF GERMPLASM INTRODUCTIONS

One of the last but not least wheat introductions of this Bicentennial period is the Japanese and Korean semidwarf germplasm brought to the United States by S. C. Salmon in 1948 (Fig. 3). One of these, 'Norin 10,' interestingly has both Fultz and Turkey wheat in its pedigree. 'Gaines' wheat was the first of a series of wheats with this germplasm.

Undoubtedly someone else compiling a list of landmark wheat introductions might include others or delete some of these, but the major ones would be common to all lists. The influence of these various germplasms has extended beyond the particular wheat class boundary. Two representative pedigrees will serve as illustrations (Fig. 4 and 5). 'Gage' wheat has in its parentage the soft red winter wheats Mediterranean and 'Kawvale,' Turkey

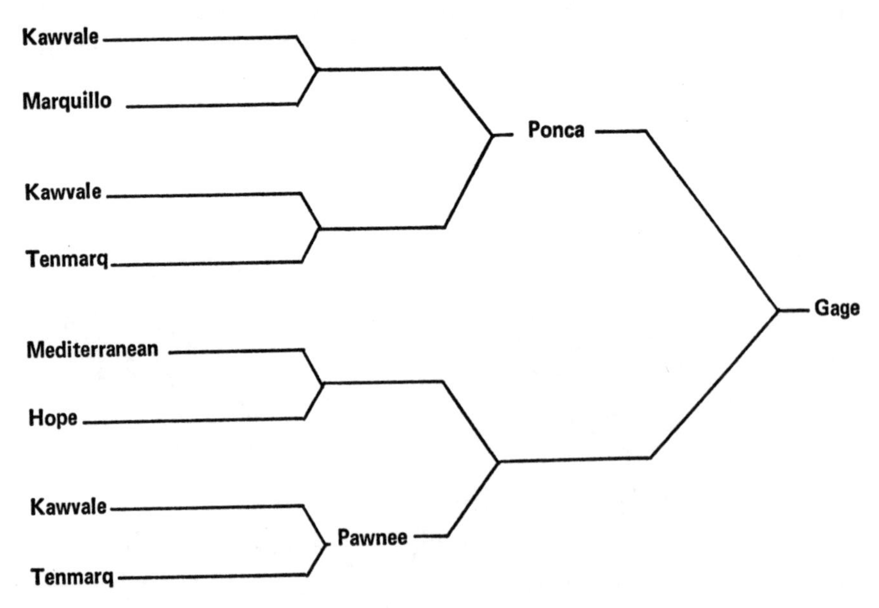

Fig. 4. Pedigree of Gage hard red winter wheat includes three wheat introductions of major importance: Marquis, Mediterranean, and Turkey.

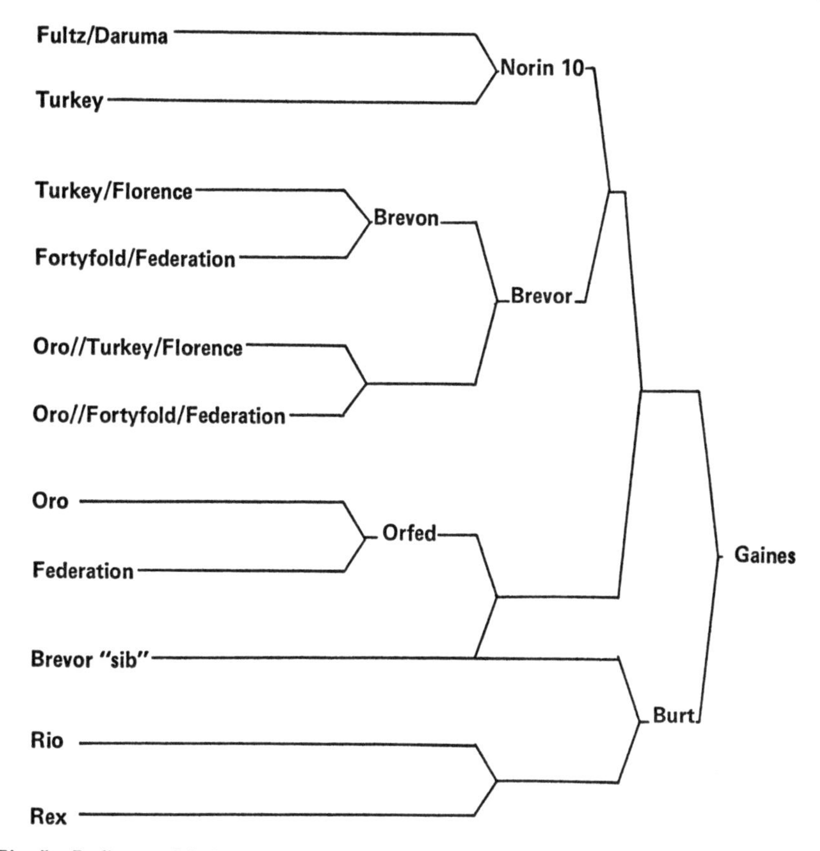

Fig. 5. Pedigree of Gaines soft white winter wheat includes three wheat introductions of major importance: Norin 10, Federation, and Turkey.

hard red winter wheat, Marquis hard red spring, 'Iumillo' spring durum through 'Marquillo,' and 'Yaroslav' emmer through 'Hope.' Gaines soft white winter wheat derived germplasm from Turkey hard red winter, 'Fortyfold' or Goldcoin (selection from Fultz?) soft white winter, Federation soft white spring, and Norin 10 soft red winter among others.

Putting it all together, what have we got? First, from this germplasm base, through selection and crossbreeding, we have the most varied array of wheat types available anywhere in the world to suit the needs of the consuming American public and for export abroad. Soft red intermediate wheats predominate in the Southeast, soft white winter wheats in New York and Michigan, soft red winters in the Cornbelt, hard red winters in the Middle and Southern Great Plains, hard red springs and amber durums in the Northern Great Plains, and white winters, both club and common, in the Pacific Northwest. Second, during the Bicentennial period we have been an "introduction"-consuming country, and while we will continue to introduce germplasm materials, the resources of our public and private breeding programs now should provide a germplasm reservoir for us and others.

LITERATURE CITED

Becker, J. A., Paul Froehlich, Donald Jackson, S. W. Mendum, F. J. Rossiter, C. V. Whalin, and Rodney Whitaker. 1941. Agricultural Statistics, 1941. USDA, Washington, D. C. 731 pp.

Briggle, L. W., and L. P. Reitz. 1963. Classification of *Triticum* species and of wheat varieties grown in the United States. Technical Bulletin 1278. USDA, ARS, Washington, D. C. 135 pp.

Carrier, Lyman. 1923. The beginnings of agriculture in America. McGraw-Hill Book Co., Inc. New York, N. Y. 323 pp.

Clark, J. Allen. 1936. Improvement in wheat. p. 207–302. *In* 1936 Yearbook of Agriculture. USDA, U. S. Government Printing Office, Washington, D. C.

Clark, J. A., J. H. Martin, and C. R. Ball. 1922. Classification of American wheat varieties. Bulletin No. 1074. USDA, Washington, D. C. 238 pp.

Clark, J. A., and B. B. Bayles. 1935. Classification of wheat varieties grown in the United States. Technical Bulletin No. 459. USDA, Washington, D. C. 164 pp.

———, and ———. 1942. Classification of wheat varieties grown in the United States in 1939. Technical Bulletin No. 795. USDA, Washington, D. C. 146 pp.

———, and ———. 1954. Classification of wheat varieties grown in the United States in 1949. Technical Bulletin No. 1083. USDA, Washington, D. C. 173 pp.

Economic Research Service. 1976. Wheat situation-238. Commodity Economic Division. USDA, Washington, D. C. 36 pp.

Edgar, W. C. 1911. The story of a grain of wheat. D. Appleton and Co. New York, N. Y. 195 pp.

Flint, C. L. 1973. The first centennial. A hundred years of progress of American agriculture. (Reprint of a paper presented to the Massachusetts Board of Agriculture at its annual meeting 1873). Agric. Sci. Rev. 11:1–17.

Quisenberry, K. S., and L. P. Reitz. 1974. Turkey wheat: The cornerstone of an empire. Agric. History XLVIII(1):98–110.

Rasmussen, W. D. 1960. Readings in the history of American agriculture. University of Illinois Press. Urbana, Ill. 340 pp.

———. 1975. Agriculture in the United States. A documentary history. Volumes 1 and 2. Random House. New York, N. Y. 1990 pp.

Reitz, L. P. 1954. Wheat breeding and our food supply. Econ. Bot. 8:251–268.

Statistical Reporting Service. 1976. Crop production. CrPr 2-2. (9–76). Crop Reporting Board. USDA, Washington, D. C. 34 pp.

Soil Physics—Reflections and Perspectives[1]

Dale Swartzendruber

INTRODUCTION

Soil physics has been defined as the study of the state and transport of matter and energy in the soil (Hillel, 1971), or as the study of the physical properties of soils (Baver, 1948). Here the word "physical" needs to be narrowly conceived as the adjective form of "physics," to avoid conveying the biological connotation as in body versus mind or spirit. Also, the second definition is somewhat circular. With even less inhibition on circularity, we could proceed as mathematicians and statisticians sometimes do (Anderson and Bancroft, 1952, p. 3), by saying for our purposes here that soil physics comprises those professional activities engaged in by soil physicists! Be that as it may, the following seven categories are listed as covering, or as being capable of stretching to cover, what it is that soil physics deals with:

1) Soil texture (particle size distribution).
2) Soil structure (particle and pore arrangement, compaction, consistency, tilth, . . .).
3) Soil water (infiltration, evaporation, transpiration and plant response, ground water, salinity and alkalinity, drainage, irrigation, . . .).
4) Soil air (exchange of oxygen and carbon dioxide).
5) Soil heat and temperature.
6) Soil erosion and stability.
7) Others (electricity, magnetism, color, acoustics, . . .).

The primary motivation of soil physics has been the dual one of obtaining a better understanding of soil physical processes to aid ultimately in the production of crops. This heavily utilitarian agricultural objective, however, has been steadily supplemented over the past 15 years by expanding interests in (i) hydrology and water resources and (ii) various environmental and pollu-

---

Dale Swartzendruber is professor of soil physics in the Department of Agronomy at Purdue University, West Lafayette, Ind.

[1]Contribution from the Department of Agronomy, Journal Paper No. 6602, Purdue Univ. Agric. Exp. Stn., West Lafayette, Ind.

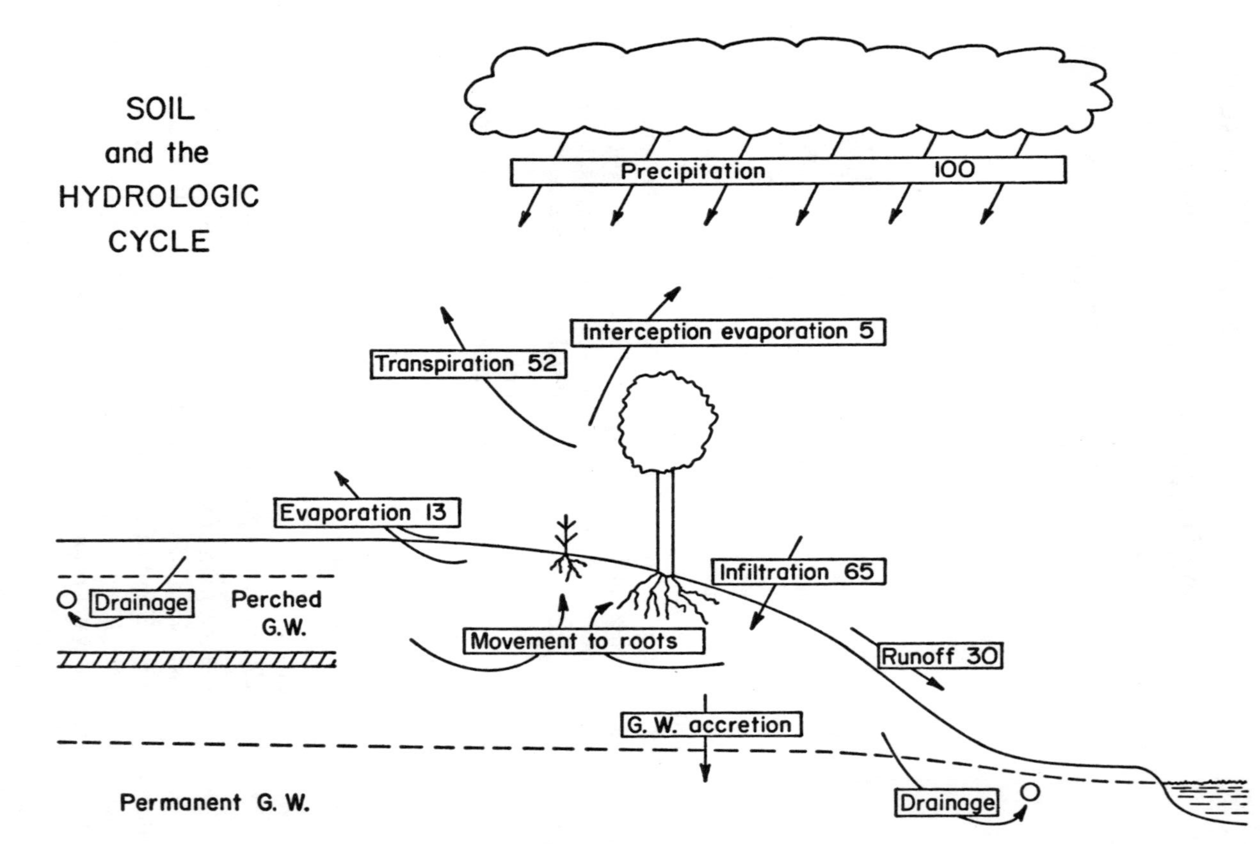

Fig. 1. Diagram of the land phase of the hydrologic cycle. Numbers in the rectangular labels are average percentages of annual precipitation in a typical humid region.

tion concerns. These expanding interests can be illustrated by the diagram of the hydrologic cycle shown in Fig. 1. It is noteworthy that a preponderance of the aspects and processes of the land phase of the hydrologic cycle are synonymous with, or closely related to, the soil-water portion of soil physics as listed above. Also, soil erosion (item 6 of list) is normally considered a concern of hydrology. Furthermore, recognizing that the hydrologic cycle is also a cycle of energy, then soil heat and temperature (item 5 of list) can likewise be included. It is thus clear that a substantial portion of the hydrologic cycle falls centrally and importantly into soil physics. Figure 1 also provides a helpful framework for observing where various concerns with pollution and environmental quality would enter in.

The numbers in the rectangular labels of Fig. 1 are percentages that partition the total annual precipitation into its various dispositions. Infiltration, in accounting for about two-thirds of the annual precipitation, represents a fairly typical value for a humid region, but shifts markedly in different climatic regions as demonstrated in Fig. 2 by the shaded portion of each total annual precipitation bar. The part of the annual precipitation available for runoff, designated by the clear portion of each bar (Fig. 2), undergoes even greater relative change than does the infiltration part.

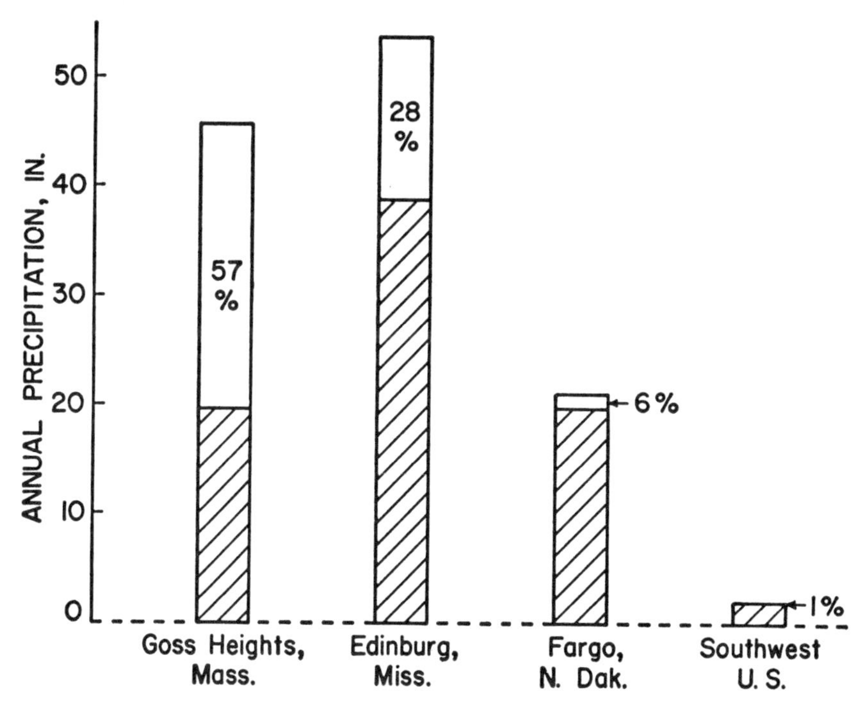

Fig. 2. Average annual precipitation (total height of bar for each location) as partitioned into infiltration (shaded portion of each bar) and runoff (clear portion), at four locations in the United States.

EARLY HISTORICAL ASPECTS OF SOIL PHYSICS

Beginnings

In the early 1700's, the English farmer and experimentalist Jethro Tull (1829) developed and presented his ideas on plants and soils. He likened the roots of plants to the intestines of animals, and suggested that the roots had "lacteal mouths" for ingesting the finest particles of soil as plant food. The soil pore space was envisioned as a "subterranean pasture" through which the roots could browse. The function of tillage was to subdivide the soil into small enough particles for the plant roots to get them into their lacteal mouths. Tull described explicitly the relationship between surface area and particle size as repeated subdivision of particles takes place, apparently being the first agriculturist to do this (Keen, 1931, p. 13). The effect of manure, he further argued, was merely to assist and promote the subdivision of the soil particles, although this could be done much better by tillage. To accomplish this tillage subdivision of the soil, Tull invented two implements. The first was the grain drill, to plant wheat and other grains in rows spaced widely enough apart for a horse to walk between. The second was the so-called horse hoe, to be drawn between the grain rows at frequent intervals for promoting particle subdivision all through the growing season.

Tull certainly had a physical concept of soil fertility and plant nutrition if there ever was one. While today his ideas may seem very strange, they did fill a crucial role in both the invention of the grain drill and the introduction of row-crop culture into England. In a scientific context, perhaps this should remind us that a theoretical explanation can seem to account for the facts and be useful in a practical sense, but still be founded on incorrect premises.

The commonly ascribed beginning of soil physics is identified with Sir Humphry Davy, the English chemist who discovered nitrous oxide (laughing gas). Definite recognition of soil physical properties appeared in his book *Elements of Agricultural Chemistry.* Davy (1813, p. 24) held high hopes for the utility of science, writing as follows:

> . . .The great purpose of chemical investigation in Agriculture ought undoubtedly to be the discovery of improved methods of cultivation. But to this end, general scientific principles and practical knowledge are alike necessary. The germs of discovery are often found in rational speculations, and industry is never so efficacious as when assisted by science.

A well developed emphasis on soil physical properties next appears in Germany, in the 1833 book *Grundsätze der Agricultur-Chemie* by G. Schübler (Baver, 1948, pp. 1-2). The origins of soil physics as a discipline are thus clearly to be found in agricultural chemistry.

Eclipse and Reemergence

Ironically, it was the continuing development of chemical viewpoints over the several decades after Schübler that almost smothered the fledgling soil-physics discipline. The overpowering influence of the chemical explanation of soil fertility was ably and eloquently advanced by J. Liebig in Germany, J. B. Boussingault in France, and J. B. Lawes and J. H. Gilbert in England (Keen, 1931, p. 35). It was during this eclipse that soil physical properties were treated for the first time in a book with a physics title, W. Schumacher's *Die Physik* in 1864 (Baver, 1948, pp. 3, 265), but its immediate influence was almost completely overshadowed by the chemical enthusiasm of the day.

Revival of interest in soil physics took place in the latter part of the 19th century, under the influence of S. W. Johnson, E. W. Hilgard, F. H. King, and M. Whitney in the United States (e.g., see W. H. Gardner, 1977) and E. Wollny in Germany. Wollny not only conducted extensive investigations in soil physics but also laid special emphasis on hydrology. In light of the recent renewal of interest in hydrologic implications of soil physics, as noted already in the present paper, it appears that hydrologic concerns have now come full circle. Another important impact of Wollny's was his editing of the journal *Forschungen auf dem Gebiete Agrikulturphysik* over the two decades 1878–98. According to Keen (1928) this was the only journal ever to be devoted primarily to soil physics. Perhaps, with the centennial anniversary of the beginning of Wollny's journal only a short time away, soil physicists might at least ponder whether the reestablishment of a journal of soil physics would be desirable for the future development of the discipline.

Whitney's emphasis on soil physics was a distinct shift from the chemical orthodoxy of the mid-nineteenth century. He placed great store on soil texture as a characterizer of soil fertility, with special stress laid on the number of particles per unit mass of soil, and on the specific surface area of the soil in relation to particle size. The yield stimulations from manures and fertilizers were viewed as brought about indirectly by their favorable effects on soil physical properties. Specifically, he wrote (Whitney, 1892, pp. 250–251):

> . . .Chemical analysis has its part to play, but we have yet to get the key to the interpretation of its results. And this key is to be found in the study of the physical structure of the soil and the physical relation to meteorology and to plant growth. . .
>
> Our work, then, is on the physical structure of the soil and its relation to the circulation of water—the movement of the rainfall after it enters the soil, and the physical effect of fertilizers and manures thereon, as related to crop production.

Reiteration and amplification of these views a decade later (Whitney and

Cameron, 1903) shows that they were not merely the initial and provisional thrusts of a beginning researcher. Such a strongly physical explanation of soil fertility had probably not been offered since the time of Jethro Tull.

FURTHER TRENDS AND DEVELOPMENT

A two-decade time interval, centered approximately on the turn of the century (i.e., 1900), can be recognized as a somewhat unusual focal period in soil physics. Identifiable with this turn-of-the-century focal period are several particular research directions and foundations, of which the salient features shall now be examined.

Particle Size Analysis

Although the sizes of soil particles had been considered and measured at least three decades before 1900, it was not until the turn-of-the-century focal period that particle size analysis (often called mechanical analysis) emerged as one of the dominant themes in soil physics. Numerous investigators employing a variety of methods were active right up to World War II. In the 1920's, a worldwide cooperative study was conducted under the auspices of Commission I of the International Society of Soil Science (Keen, 1928, 1930). Size ranges for defining the sand, silt, and clay contents of soils were designated, and became the basis throughout soil science for identifying and classifying the texture of soils. In turn, the texture became an important element in the overall classification of soils.

An interesting use of particle size analysis was being developed during the turn-of-the-century focal period by J. Kopecký in Bohemia (Czechoslovakia). Having designed a serial elutriator (Baver, 1948, pp. 42, 61) for separating soil particles into consecutive size classes, Kopecký then studied the spacing of tile drains as related primarily to the content of particles < 0.01 mm and secondarily to the particle content in the size range 0.01–0.05 mm. This relationship of drain spacing to particle size was expressed and summarized in tables and nomograms that became the prevailing basis for the design of tile drainage systems in Czechoslovakia and central Europe, continuing up to World War II. An illustration is given in Table 1, which was kindly supplied for use here by Dr. Pavel Dvořák, Department of Irrigation and Drainage, Technical University, Prague. It is recognized that, physically, particle size classes might seem to be too indirect to provide an adequate characterization of soil drainability. Nevertheless, this was indeed a soil physical approach to soil drainage, which, for its time in history, merits recognition for the way it was used in practical drainage design. Furthermore, even today we still await a practicable drain-spacing prescription based on a transport coefficient such as the saturated soil hydraulic conductivity. Part of the difficulty is that of obtaining a relevant measurement of the hy-

Table 1. Tile-drain spacing in relation to soil particle size, as developed by J. Kopecký.†

Qualitative soil permeability	Particle content in size classes: <0.01 mm	0.01-0.05 mm	Tile spacing
	%	%	meters
Low	>70	>55	8-9
	70-55	55-40	9-10
	55-40	40-25	10-12
Moderately permeable	40-30	25-15	12-14
	30-20	15-7	14-16
	20-10	7-2	16-18
Permeable	<10	<2	18-20

† Data of 1911, courtesy of Dr. P. Dvořák, Technical University, Prague, Czechoslovakia.

draulic conductivity itself. Hence, more historical awareness of Kopecký's drain-spacing work would appear to be very much in order.

Classification of Soil Water

Following the prevailing ideas at that time, Briggs (1897) proposed that soil water should be classified as gravitational, capillary, and hygroscopic. Great interest and activity in soil physics were then directed along these lines that had somehow been so effectively articulated by Briggs. Considerable effort was expended in further subdivision and refinement of the classification scheme, so that by 1930 Zunker (Baver, 1948, p. 201) had expanded the total number of subdivisions from 3 to 10, as shown in Table 2. Other efforts sought to refine and quantify the various regions and points along the water-content range, from water saturation to complete or near-complete dryness. This led to the introduction and use of such terms as field capacity and wilting point, with the so-called plant-available water taken as the difference between these two quantities. The relative simplicity of these ideas

Table 2. Summary of a detailed classification of soil water by Zunker in 1930 (Baver, 1948, p. 201).

1. OSMOTIC (organic cells)
2. HYGROSCOPIC†
3. CAPILLARY†
4. "HELD" (HAFTWASSER)
   a) film
   b) "pore angle"
   c) capillary "held"
5. GRAVITATIONAL†
   a) capillary
   b) downward
6. GROUND WATER
7. WATER VAPOR

† Briggs' (1897) three simple subdivisions.

seems to have invested them with a pervasive acceptance that persisted up to World War II and even to some degree into the present time.

Viewed in historical retrospect, the heavy classificational emphasis in studies of both particle size and soil water looks like a kind of taxonomic syndrome in much of the soil physics of the early 20th century. That is, the soil particles were size-classified into sand, silt, and clay, which were then classified into textural groupings that in turn were used in classifying soils in an overall pedological scheme. Soil-water study consisted of specifying water forms and classifying them. Such a taxonomic soil physical syndrome is not necessarily surprising, however, if we note that the late 19th and early 20th centuries were the flourishing times of the Russian soil classificationists V. V. Dokuchaev and K. D. Glinka, along with their American counterpart and expositor, C. F. Marbut (Jenny, 1961, pp. 73–75). A taxonomic touch in a physicist is revealed by W. Gardner (1936, p. 384). In consideration of simple equations for expressing soil-water suction as a function of water content, he suggested that the parameters of such simple equations would likely be suitable for a basis of soil classification.

Mathematical Directions

We now consider a matter of which many soil and crop scientists may be well aware even if their particular specialties are far removed from soil physics. Posed as a question, this would be: why do many present-day soil physicists expend so much time and effort with mathematics and equations? However, before answering the why of this question let us first consider several preliminary queries.

1) Is the substance of the posed question really true?
2) If true, when did it begin?
3) If true, are soil physicists the only soil scientists so affected?

To answer these three queries, inspection was made of the papers published in the *Soil Science Society of America Proceedings (Journal)* under the headings of Division S-1 (Soil Physics) and Division S-2 (Soil Chemistry). It was felt appropriate to include Division S-2 because of the early soil-physics origins in agricultural chemistry as already discussed. It is not to be implied that soil physicists and chemists do not publish in other journals; rather, this restricted choice of journal was made solely to keep the inspection task within reasonable bounds. For the same reason, only about every fifth year (volume) was analyzed, beginning with the first volume of the *SSSA Proceedings* in 1936. For each volume analyzed, the numbers of pages, papers, and mathematical equations (but not chemical equations) were determined in Divisions S-1 and S-2, along with the total numbers of papers and pages in the complete volume.

The number of pages per year is plotted against calendar year in Fig. 3. The publication activity in all soils divisions increased steadily over the 40-

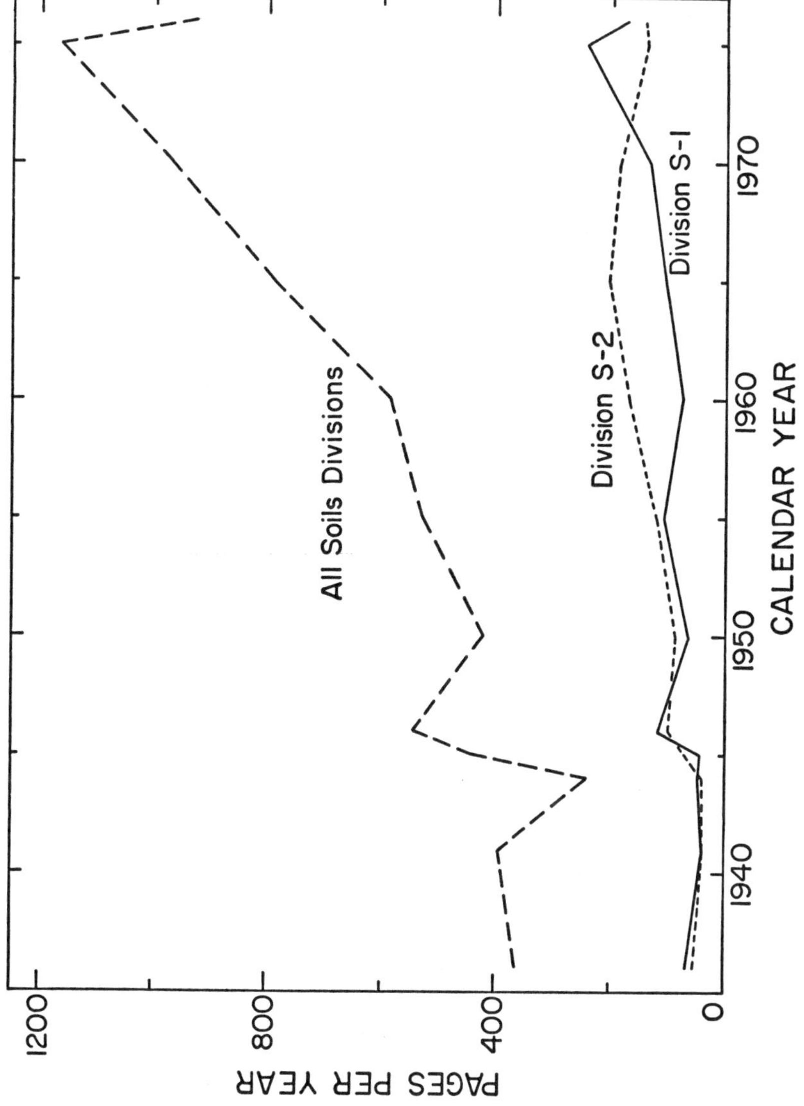

Fig. 3. Time course of annual number of pages of papers published in the *SSSA Proceedings (Journal)* in Division S-1 (Soil Physics), Division S-2 (Soil Chemistry), and all Divisions.

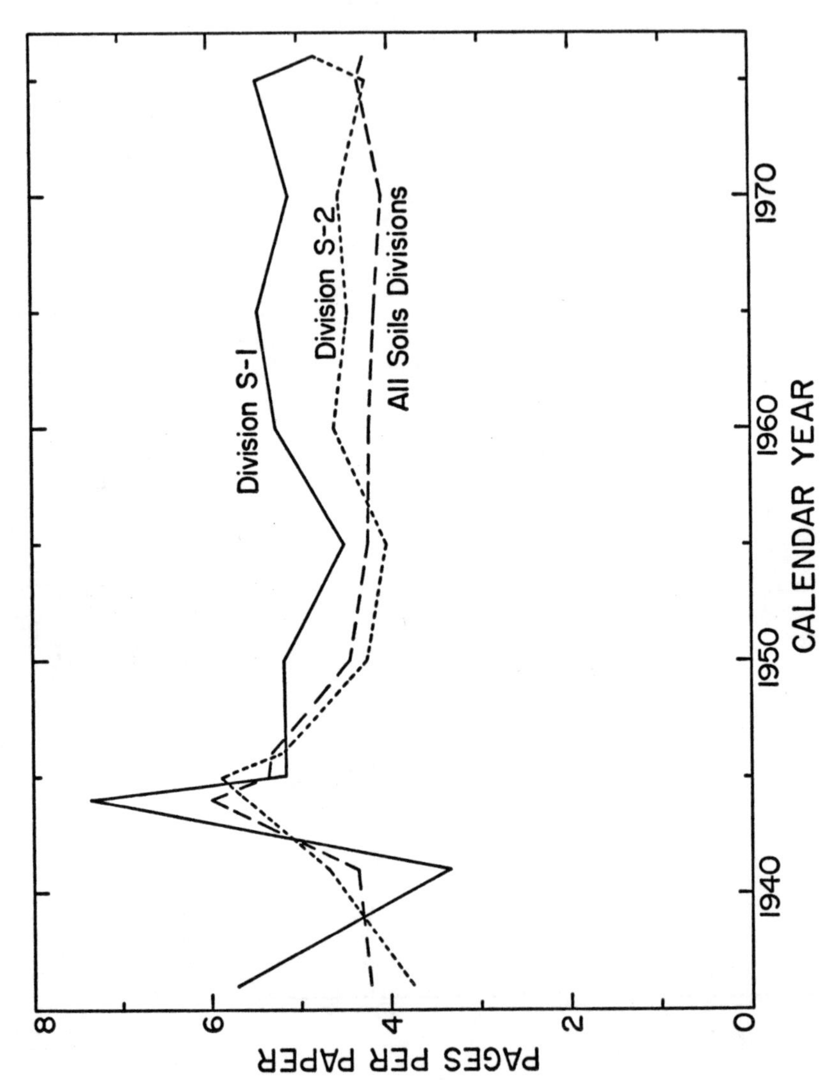

Fig. 4. Time course of number of pages per paper, for papers published in the *SSSA Proceedings (Journal)* in Division S-1 (Soil Physics), Division S-2 (Soil Chemistry), and all Divisions

year period, except for a dip and peak associated with World War II. Activities in Divisions S-1 and S-2 were fairly similar to each other, with very close correspondence up to 1955, but with somewhat differing trends thereafter. The trends in both Divisions follow similarly to that of the total Society (all soils divisions).

When the number of papers per year was plotted against calendar year, the behavior picture was found to be very close to that of Fig. 3, and hence is not shown here. Instead, the number of pages per paper is plotted against calendar year in Fig. 4, where the general trends are fairly constant in S-1, S-2, and overall, except for some fluctuations up to and through World War II. After 1960, the number of pages per paper averaged slightly above 5 in S-1, about 4.5 in S-2, and very slightly above 4 in all soils divisions. Hence, the overall Society has seemingly adhered closely to the 4-page free publication limit.

With Figs. 3 and 4 as background, we next turn our attention to the number of equations per page as plotted in Fig. 5 for Divisions S-1 and S-2. This clearly answers our three preliminary queries. First, the use of mathematics by soil physicists, as expressed by the number of equations per page appearing in S-1 papers, has indeed shown a marked and strong increase over the past 30 years. Second, this began near the end of World War II, from essentially no equations both then and prior to the War. Third, soil physicists are not alone in the increased use of mathematics, since, beginning also near the end of World War II, the number of equations per page in S-2 has increased steadily from near zero, although the magnitude and rate of increase are not as marked as in S-1. The increase of equations in S-1 since 1965 occurs in concert with an increase of S-1 pages (Fig. 3). Finally, we shall note in Fig. 5 that the sharp peak of the S-1 curve in 1945 could be called the Kirkham spike, because of the dominating role of a single paper (Kirkham, 1945).

Let us now deal with the questions posed originally—why mathematical activity is so definite in soil physics. The reasons begin with the flux equation introduced by Darcy (1856) for water flow through water-saturated sand. Shown in Fig. 6 are the illustrative flow system and his equation, which is simply a quantitative statement of water seeking its lowest level by seeping through the porous medium (sand) characterizable by the hydraulic conductivity constant K. Mathematically, Darcy's equation is of the same form as Newton's law of cooling (heat flow) and Ohm's law (flow of electricity).

If the principle of mass balance (or conservation of mass) is combined with Darcy's equation, as done very generally by Slichter (1898), the result is the classic Laplace equation that is the basis of a very powerful and general flow-problem-solving framework. Summarizing these broad steps of the development of the framework (for water-saturated flow):

1) Darcy equation: flux = (hyd. cond.) $\times$ gradient
2) Mass balance (or continuity) equation:
   soil-water storage = water inflow minus water outflow

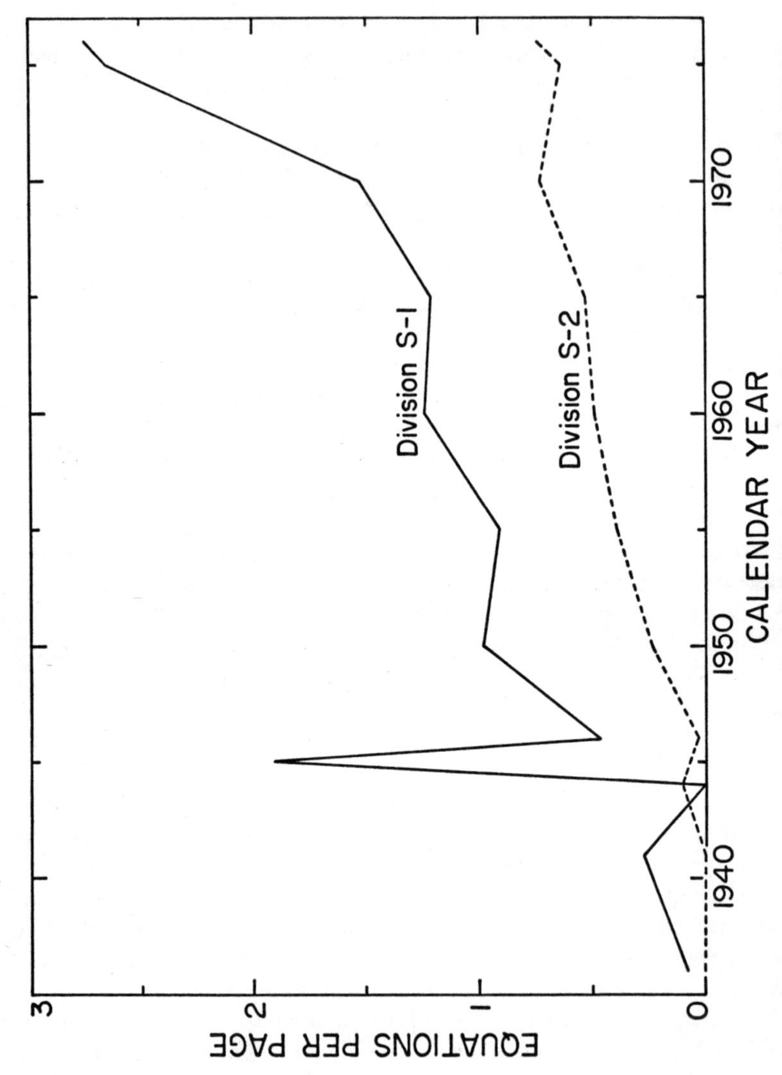

Fig. 5. Time course of number of mathematical equations per page, for papers published in the *SSSA Proceedings (Journal)* in Division S-1 (Soil Physics) and Division S-2 (Soil Chemistry).

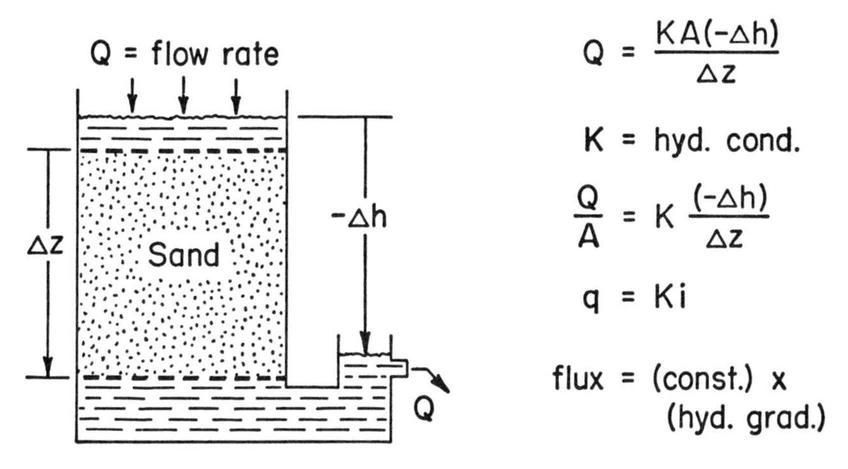

Fig. 6. Flow system and Darcy's (1856) equation for the flux of water through a water-saturated sand column of cross sectional area A and hydraulic conductivity K.

3) Combine (1) and (2) (e.g., for z and x directions):

$$\frac{\partial^2 h}{\partial x^2} + \frac{\partial^2 h}{\partial z^2} = 0 \qquad \text{(Laplace equation)}$$

The solution *in principle* for a saturated-flow problem is to solve the Laplace equation together with the boundary equations. Such a conceptual and problem-solving tool is very useful. Even so, it took some several decades for the tool to come into appreciable use among soil physicists and others dealing with soil drainage.

The general condition of water in soil is obviously not that of the soil pores always being saturated. To describe unsaturated flow, Buckingham (1907) introduced the two soil-characterizing functions illustrated graphically in Fig. 7. $K = K(\theta)$ is the hydraulic conductivity function. The quantity $\tau = \tau(\theta)$ represents what is presently called the soil-water or matric suction (earlier, soil-moisture tension), and which was first identified by W. Gardner et al. (1922) as measurable by a porous-element tensiometer. With no reference whatever to Darcy, but arguing from analogy with the flow of heat and electricity, Buckingham then postulated a proportional flux equation and inserted into it his two water-content functions, $K(\theta)$ and $\tau(\theta)$, as indicated in the bottom part of Fig. 7. To rectify his omission of reference to Darcy, however, we shall here refer to his flux relationship as the Buckingham-Darcy equation (Swartzendruber, 1966, 1969).

If the principle of mass balance is combined with the Buckingham-Darcy equation, as was first done in the most general form by Richards (1931), the result is now commonly called the Richards equation. Summarizing the broad steps of the development of this flow-problem-solving framework (for unsaturated flow):

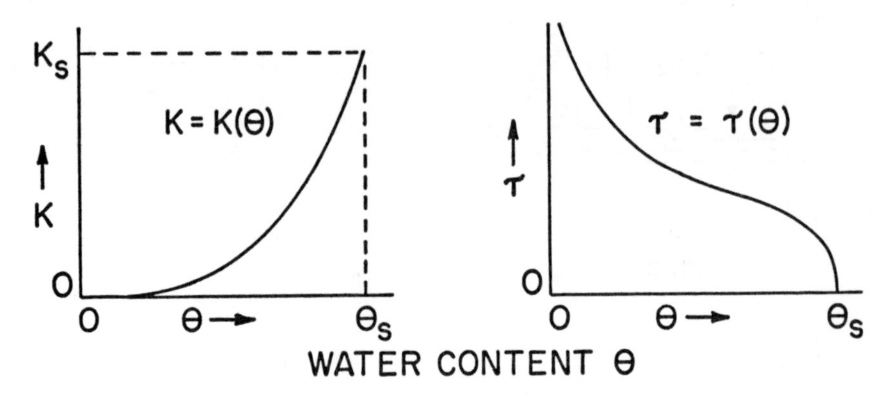

flux q = Ki = K(-Δh/Δz) as before, but:

K = K(Θ), and τ = τ(Θ) in Δh = Δz - Δτ

Fig. 7. Buckingham's (1907) functions, $K = K(\theta)$ and $\tau = \tau(\theta)$, for characterizing water flow in an unsaturated soil or porous medium, where $q = Ki$ is the Buckingham-Darcy flux equation (Swartzendruber, 1966, 1969).

1) Buckingham-Darcy equation:
   flux = (hyd. cond. function) X gradient
2) Mass balance (or continuity) equation:
   soil-water storage = water inflow minus water outflow
3) Combine (1) and (2) (e.g., for direction z and time t):

$$\frac{\partial}{\partial z}\left[K(\theta)\frac{\partial \tau}{\partial z}\right] - \frac{\partial K(\theta)}{\partial z} = -\frac{\partial \theta}{\partial t} \qquad \text{(Richards equation)}$$

The solution *in principle* for an unsaturated-flow problem is to solve the Richards equation together with the boundary and initial equations. Once again, the appeal and utility of such a conceptual and problem-solving tool for unsaturated-flow analysis is obvious.

The use of mathematics in soil physics, as reflected in Fig. 5, is thus a logical outcome of the problem-solving frameworks provided by the Laplace and Richards equations. Note that two of the crucial contributions to these frameworks, Slichter (1898) and Buckingham (1907), occurred during the turn-of-the-century focal period already mentioned in this paper.

An ironic reversal has occurred during the long period between Buckingham's (1907) work and its eventual fruition in mathematical unsaturated-flow theory following World War II. We have already noted that Buckingham made no mention of Darcy. But, over recent decades and even at present, most soil-water physicists do not use Buckingham's name when referring to the flux equation for unsaturated flow. Instead, they will likely state something about applying Darcy's equation to unsaturated flow. Of six recent textbooks, two (Childs, 1969; Nielsen et al., 1972) make no mention whatever of Buckingham, while the other four (Hillel, 1971; Baver et al., 1972; Kirkham and Powers, 1972; Taylor and Ashcroft, 1972) make only oblique or passing reference. Specific recommendations for using Buckingham's name have been offered (Richards, 1960; Swartzendruber, 1966, 1969), but seemingly with little impact. Perhaps poetic justice is being meted out for Buckingham's original failure to mention Darcy!

Simplified Analysis of Soil-Water Movement

At the end of the turn-of-the-century focal period, Green and Ampt (1911) introduced an approximate soil-water flow analysis that has found a recent renewal of interest. Soil-water movement is treated as though there were an abrupt front between two regions of soil at different water contents. As illustrated specifically for a water infiltration case in Fig. 8, the rectangular "Green-and-Ampt" profile replaces the actual and more gradual profile of water content. This yields a distinct reduction in mathematical complexity, but still provides fairly accurate characterizations of such quantities as the overall rate and cumulative volume of water transport. The approach is not applicable for quantitative description of the soil-water distribution.

Using this approach to describe the runoff from a small-plot sprinkling infiltrometer, measurements of saturated hydraulic conductivity have been obtained in the field on areas of about 1 m^2 (K. A. Sulaiman, 1976. Saturated soil hydraulic conductivity as determined in the field with a small-plot

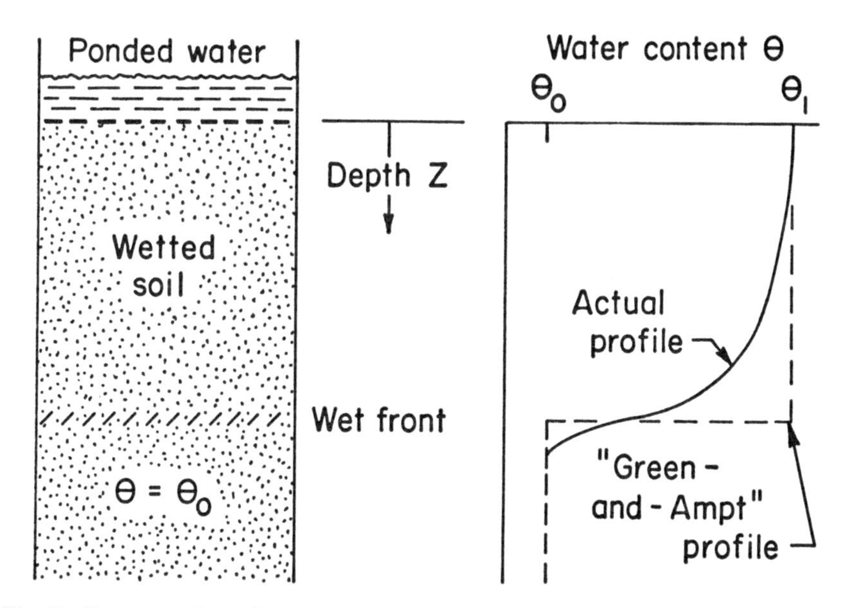

Fig. 8. Representation of a water-content profile by the rectangular approximation of Green and Ampt (1911), as illustrated for water infiltration from a ponded soil surface.

sprinkling infiltrometer. M.S. Thesis. Purdue University, West Lafayette, Ind.). For such repeated measurements in close proximity in the field, the ratio of largest value to smallest value was only 1.2. This compares with a ratio of 155 for so-called undisturbed soil cores and with ratios of 2.1 to 20.7 for soil-cavity field methods (Reeve and Kirkham, 1951). Such encouraging comparisons give hope that meaningful soil-water transport parameters can indeed be evaluated for field conditions.

OTHER CONSIDERATIONS

The individual research paper is a vital building block of information and knowledge in any field. The integration of knowledge and the development of a discipline, however, would also seem to require the preparation of monographs and books, especially for educational and training purposes. It was thus deemed of interest to enumerate the soil physics books published through the years. Selection criteria for these books were not always easy to prescribe, but it is felt that a reasonable consistency was still attainable. The results are shown in Fig. 9, plotted as cumulative number of books against time, beginning with Davy's book in 1813. The curve in its early

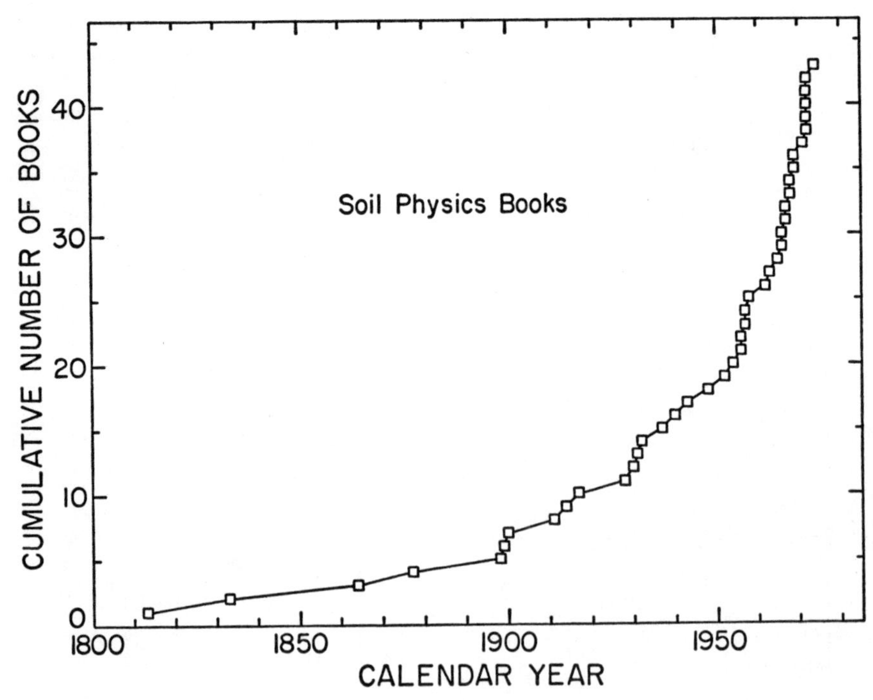

Fig. 9. Time course of the cumulative number of soil physics books, beginning with that of Davy (1813).

part is rather flat, but begins to climb more strongly upward during the turn-of-the-century focal period. Within the past decade (1966–76) the curve rises quite steeply.

If the number of soil physics books could be taken to reflect the general activity and number of scientists in the discipline, then the graph of Fig. 9 would also represent a kind of soil-physics growth curve. The curve does seem compatible with the time graph of W. H. Gardner (1977) for the number of soil physicists in the United States. It is also of interest to compare the curve with Philip's (1974) estimates, that 90% of all work ever done in soil physics was done in the last half century, and that 80% of all soil physicists that ever lived are alive today.

FUTURE OUTLOOK

Prediction of the future is a risky exercise, and I am always wary of attempts to do so. Nevertheless, it seems that there are and will be numerous opportunities for soil physics to make vital contributions toward solving the problems of food, water, and the environment. In developed countries some of these would be

1) more efficient use of soil water by crops, in the total context of yield modeling;
2) soil structure and compaction problems;
3) behavior and interaction of water, salts, fertilizers, and pesticides in the soil;
4) soil drainage, humid and arid (under irrigation);
5) soil erosion and sediments; and
6) waste disposal on land, and the effects on soils.

While the foregoing considerations will apply also in developing countries, even special urgencies are there involved. A recent report (CIMMYT, 1976) points out that of the 30% increase in grain production in developing countries during the past decade, *only half* came from increased yields per hectare; the other half came from increasing the cultivated land area. So, we are still a long way from achieving the many-fold yield increases inherently resident in the new crop varieties of the "Green Revolution." Hence, greatly improved management of both physical and chemical soil fertility is still needed. To draw a comparison with mathematics, it would seem that an improved crop variety is indeed a *necessary* condition for a broad-scale yield-per-hectare increase, but may well not be a *sufficient* condition, the sufficiency having to do rather crucially and vitally with the physical, chemical, economic, and other aspects of the management "package." Finally, in a Peace Corps brochure on agronomy (*Agronomy*, GPO 940-534, Action Flyer 4200-3, July 1972), the question was posed as to how tropical soils would react to intensive agriculture, more specifically:

1) Will they erode?
2) Will they compact?

3) What about micronutrients?
4) Will chemical fertilizers leach away too rapidly for maximum uptake by cultivated plants?

Clearly, three out of four of these questions are heavily soil physical.

It is my feeling that soil physics presently stands at a stage of great opportunity and challenge. To be sure, there is much unfinished business, such as just getting underway with the crucial tasks of testing and validating mathematical-physical theories under field conditions, but definite progress is being made. Where gaps exist, whether between principles and practice within soil physics or between soil physics and other related areas, such gaps can be bridged by mounting concerted efforts from both sides of each gap, rather than expecting an effort from one side to go all the way. Vital to all efforts will be the adequate provision of financial and other resources for the training of students and for supporting them in their subsequent careers.

In this historic and commemorative year, let us renew our sense of dedication and purpose, to the end that our urgent problems of food, water, and the environment will be solved adequately. May we work diligently that the world might indeed witness the attainment of the worthy expectations of Sir Humphry Davy (1813, pp. 26–27) when he wrote:

> Discoveries made in the cultivation of the earth are not merely for the time and country in which they are developed, but they may be considered as extending to future ages, and as ultimately tending to benefit the whole human race, as affording subsistence for generations yet to come, as multiplying life, and not only multiplying life but likewise providing for its enjoyment.

LITERATURE CITED

Anderson, R. L., and T. A. Bancroft. 1952. Statistical theory in research. McGraw-Hill Book Co., Inc., New York.

Baver, L. D. 1948. Soil physics. 2nd ed. John Wiley and Sons, Inc., New York.

Baver, L. D., W. H. Gardner, and W. R. Gardner. 1972. Soil physics. 4th ed. John Wiley and Sons, Inc., New York.

Briggs, L. J. 1897. The mechanics of soil moisture. USDA Weather Bur. Div. Soils, Bull. 10.

Buckingham, E. 1907. Studies on the movement of soil moisture. USDA Bur. Soils, Bull. 38.

Childs, E. C. 1969. The physical basis of soil water phenomena. John Wiley and Sons Ltd., London.

CIMMYT, Centro Internacional de Mejoramiento de Maiz y Trigo. 1976. CIMMYT review 1976. International Maize and Wheat Improvement Center, El Batan, Mexico.

Darcy, H. 1856. Les Fontaines Publique de la Ville de Dijon. Victor Dalmont, Paris.

Davy, H. 1813. Elements of agricultural chemistry. Longman, Hurst, Rees, Orme, and Brown, London.

Gardner, W. H. 1977. Historical highlights in American soil physics, 1776-1976. Soil Sci. Soc. Am. J. 41:221-229.

Gardner, W. 1936. The influence of soil characteristics on drainage and irrigation practices. Soil Sci. Soc. Am. Proc. 1:383-392.

Gardner, W., O. W. Israelsen, N. E. Edlefsen, and H. Clyde. 1922. The capillary potential function and its relation to irrigation practice. Phys. Rev. (Series 2) 20:196.

Green, W. H., and G. A. Ampt. 1911. Studies on soil physics: 1. The flow of air and water through soils. J. Agric. Sci. 4:1-24.

Hillel, D. 1971. Soil and water: Physical principles and processes. Academic Press, Inc., New York and London.

Jenny, H. 1961. E. W. Hilgard and the birth of modern soil science. 3 Collana Della Rivista "Agrochimica", Pisa, Italy.

Keen, B. A. 1928. First Commission. Soil mechanics and physics. Soil Sci. 25:9-20.

Keen, B. A. 1930. Preliminary report on "single-value" cooperative work. Int. Congr. Soil Sci. Proc. 2nd (Leningrad), Commission I (Soil Physics), p. 1-7.

Keen, B. A. 1931. The physical properties of the soil. Longmans, Green and Co., London.

Kirkham, D. 1945. Proposed method for field measurement of permeability of soil below the water table. Soil Sci. Soc. Am. Proc. 10:58-68.

Kirkham, D., and W. L. Powers. 1972. Advanced soil physics. John Wiley and Sons, Inc., New York.

Nielsen, D. R., R. D. Jackson, J. W. Cary, and D. D. Evans, ed. 1972. Soil water. Am. Soc. Agron., Madison, Wis.

Philip, J. R. 1974. Fifty years progress in soil physics. Geoderma 12:265-280.

Reeve, R. C., and D. Kirkham. 1951. Soil anisotropy and some field methods for measuring permeability. Trans. Am. Geophys. Union 32:582-590.

Richards, L. A. 1931. Capillary conduction of liquids through porous mediums. Physics 1:318-333.

Richards, L. A. 1960. Advances in soil physics. Int. Congr. Soil Sci. Trans. 7th (Madison, Wis.) I:67-79.

Slichter, C. S. 1898. Theoretical investigation of the motion of ground water. U. S. Geol. Survey 19th Annual Report, part 2, p. 295-384.

Swartzendruber, D. 1966. Soil-water behavior as described by transport coefficients and functions. Adv. Agron. 18:327-370.

Swartzendruber, D. 1969. The flow of water in unsaturated soils. p. 215-292. *In* R. J. M. DeWiest (ed.) Flow through porous media. Academic Press, Inc., New York.

Taylor, S. A., and G. L. Ashcroft. 1972. Physical edaphology. W. H. Freeman and Co., San Francisco.

Tull, J. 1829. The horse-hoeing husbandry. William Cobbett, London.

Whitney, M. 1892. Soil investigations. Maryland Agric. Exp. Stn., 4th Annual Report 1891, p. 249-296.

Whitney, M., and F. K. Cameron. 1903. The chemistry of the soil as related to crop production. USDA Bur. Soils, Bull. 22.

Focus on the Future with an Eye to the Past

William F. Hueg, Jr.

During the Bicentennial Year the American public paused long enough from its daily tasks to give recognition to where and what we have been, as a people and a nation. We have also given thought to where and what we might be as we celebrate our Tricentennial in 2076.

The Bicentennial celebrates a revolution for independence from a dominating foreign power. In 1975, when we celebrated the Centennial of the State Agricultural Experiment Station system we called attention to a peaceful revolution which freed the United States from the specter of hunger and starvation.

But around the world in the developing nations we have seen the seeds of revolution, if not actual revolution, because of the lack of food. Many of these governments, both new and old, have failed to meet the most basic need of their people–food.

American agriculture is the envy of the developing world and a paradox in the United States. In 1972, when on a world basis there was approximately a 3% reduction in world grain supplies, all stops were pulled to make it possible for American agriculture to meet its full productive potential. Already predictions are being made of a return to the "surplus days" of the late 1960's and early 1970's. Some reporters have been so bold as to suggest that because of improved harvests this year in the U.S.S.R. and much of Asia, the "world food crisis" of a few years ago has disappeared.

At the same time, many national reports call attention to the need for expanded research and education in agriculture and related areas. William Wampler, Congressman from Virginia, introduced in the 1976 session of Congress a far-reaching bill to rapidly increase the funding of research in agriculture and related fields. It did not pass but continuing interest is strong. A major amendment, Title XII, was made to the Foreign Assistance Act in 1975 and is focused on world food production. It is euphemistically referred to as the "Famine Prevention Act." The important Board of International Food and Agricultural Development is now getting organized and hopefully we will

---

William F. Hueg, Jr. is Professor of Agronomy, and Deputy Vice President and Dean of the Institute of Agriculture, Forestry and Home Economics at the University of Minnesota, St. Paul, Minn.

see a new approach to agricultural development work from this legislation and activity.

All of these events would suggest that American agriculture is on the move in its domestic development and in meeting its opportunities and responsibilities to the hungry world. Those of us close to the scene know that this is not necessarily the case.

Simply stated, American agriculture must continue to stand on its own ground and make its decisions in terms of production and marketing goals and the methods of accomplishment. The political rhetoric of an election year could have, if we had let it, lulled us into a state of euphoria. The great challenge to the Congress—and to American agriculture—is whether we can develop a national food policy in this next year which will make it possible to meet the challenges for food in the remainder of this century at home and throughout the world. It is my concern that we will only concentrate on a new agricultural bill and this may not be sufficient in scope and opportunity for developing a policy related to food and fiber so necessary for the remainder of this century.

With the tremendous land and climatic resources available to us, and the willingness and skill of the American farmer and the attendant input and output industries which serve him, production agriculture will always be important in the United States. Whether production agriculture receives the attention on the political scene that makes it possible for it to meet its full potential can only be determined by the action which the agricultural community is willing to take.

The greatness of American agriculture is not an accident, but I am not sure that we can conclude it was a grand design which has developed over the past hundred or more years. We do know that as the United States moves into the third century, American agriculture has had a vital and substantial role in all facets of our nation's development.

The growth and progress of American agriculture has been the base of an American social and economic system. The stability of our government and economy even in its times of crisis were maintained because we had relatively abundant food. The strength of American agriculture is the system which has developed. This is a system of production on the land, supplemented by and in concert with the supply industry on the one side and the processing, transporting, and distribution industry on the other. This is American agriculture—this is agribusiness!

But what about this grand design? Was there such a thing, and if so, what was it?

Newspapers, books, and speeches in the 1840's and 1850's in the United States carried such statements as "too little land, too many people, too little food!" For a country that has the abundance of today, statements of this type are hard to believe. They sound more like present day statements from developing nations around the world. But the United States, in the 1840's and 1850's, was a developing nation.

Most of our population was on the East Coast in the original 13 states.

Those who had the pioneering spirit crossed the Allegheny and Appalachian Mountains through the tortuous valleys, rivers, and streams. They took the risk; they wanted open space, and open space was something the United States had in abundance. But how to develop it? How to get people to move to the vast wilderness in the west?

And then came the grand design, now probably developed beyond the fondest dreams of the originators. I refer to it as the three significant events of 1862, all acts signed into law by President Abraham Lincoln.

SIGNIFICANT EVENTS—1862

The Homestead Act was the first of these significant events, making it possible for the vast wilderness to be opened up for settlement by those willing to take the risk of leaving the relative comfort of their homes in the more established parts of the United States and moving to the West. The government provided 64.8 ha (160 acres) of land to those willing to "prove the land" by clearing 2 ha (5 acres), building a house and having agricultural development within a period of five years. This was the beginning of the family farm as we think of it through our agricultural history; it was the beginning of a major change in our system of food production in the United States, from one of subsistence to one of production for markets. It was the development of commercial agriculture.

The Morrill Land-Grant Act was the second significant event and made it possible for every state to develop a college that would serve all of the people, but especially the sons and daughters of farmers and mechanics. Tracts of land, which could be sold or used as a base of income generation to establish these colleges, were given to the states. The act has caused a major revolution in higher education, and we find many developing nations emulating our land-grant university system.

Creation of the U. S. Department of Agriculture (USDA) was the third significant event and provided an agency within the federal government to deal directly with the problems of the people on the land. Since its inception, the responsibilities of the USDA have broadened significantly and now closely touch on consumer issues as well as those of production agriculture.

Four additional acts rounded out those of 1862 in the agricultural development of the United States. In 1887, the Hatch Act made it possible for states to establish research units as a part of their land-grant universities. The Smith-Lever Act of 1914 provided federal funding for extension services with primary emphasis on agriculture, although their responsibilities have broadened over the years. Also in 1914, the Smith-Hughes Act established vocational education essentially in the areas of agriculture and home economics. Again this has broadened considerably.

In 1922, the Capper-Volstad Act established the basis by which farmer cooperatives could be formed to market agricultural products and bargain for the selling price of products. In 1935, the Soil Conservation Service

(SCS) was established as an agency within the USDA to provide technical assistance to Soil and Water Conservation Districts established in each state on county wide or area basis.

All of these acts, then, are part of the grand design which has helped to develop the agricultural system as we know it in the United States today. However, if farmers and industry had not been willing to accept the results of research as the information was brought to them through university and industry extension programs, we would not have had the system develop as we know it today. The structure within agriculture seems cumbersome at times, but I believe the diversity of organization to be one of its great strengths.

AMERICAN AGRICULTURE—1976

Agriculture is that industry which deals with food and fiber production from the time the land is prepared and the seeds planted until the products of crops and livestock reach the consumer. A more modern term is agribusiness. This includes farm production and the supply and processing industries so essential for the agriculture we know in America.

In 1976, there were approximately 2.8 million farms in the United States with 800,000 of these producing about 88% of the food and fiber. These farms are capitalized at over a half a trillion dollars a year, about three-fifths of the capital assets of all U. S. corporations. By comparison, the automobile industry is capitalized at about $80 billion annually.

Agriculture is the major industry of the United States and the world; it's a major consumer of many of the essential manufactured products, such as fuel, steel, rubber, and fertilizer. The food system affects each American every day, and around the world one-quarter of what the world's population eats comes from the American farm. The American farmer is part of the most elite group of our world population, making up one-tenth of one percent of that population but producing 25% of the world's food supplies.

Employment in agriculturally-related enterprises makes up 28% of the nation's labor force. Four percent of the nation's working force are on farms and for each farm worker, seven others provide backup in service, research, education, marketing, processing, and distribution. The present ratio of 1 to 7 will change to 1 to 10 or 12 by the year 2000.

Crop agriculture in 1976 produced 152.5 million metric tons (6.0 billion bushels) of corn (*Zea mays* L.); 32.7 million metric tons (1.2 billion bushels) of soybeans (*Glycine max* L. Merr.); and 57.2 million metric tons (2.1 billion bushels) of wheat (*Triticum aestivum* L.). Each year we export about 25% of the corn produced, 35% of the soybeans, and 50% of the wheat. The export value of agricultural products in 1976 will exceed $22 billion, a major factor in the positive balance of payments of the United States. Food and feed grains are a substantial part of these exports.

With the world's population nearly doubling in the remainder of this century, the challenge to agronomists and other agricultural scientists is greater than it has ever been before. Nature still holds tightly to her secrets of life, and it is only with the strongest thrusts of basic and applied science that we are able to wrench these secrets from her bosom.

If the world is to accommodate 7 billion people by the year 2000, we must ask ourselves whether as a nation we will be expected to contribute to world food supplies in the same proportion as we did in 1976. If the answer is yes, then that means our challenge in the United States is the production of 228.8 to 254.2 million metric tons (9 to 10 billion bushels) of corn, 68.1 million metric tons (2.5 billion bushels) of soybeans, and 109.0 million metric tons (4 billion bushels) of wheat. This of course, assumes that the rate of consumption will stay as it is today; that is not necessarily an optimistic prospect for millions of people who are now at low levels of food consumption and nutrition.

What does previous agronomic research and experience tell us about future opportunities to meet world food demand? I have drawn on some examples with the assistance of Minnesota colleagues, Vernon Cardwell and Robert Heiner of the Department of Agronomy and Plant Genetics and Robert Gast, Curt Overdahl, and William Fenster of the Department of Soil Science. I believe the tables and charts we have developed give reason for hope and tremendous excitement in meeting the food challenge of these next two decades. This encouragement is not only for domestic production, but suggests even greater opportunity for the less developed countries (LDC).

LONG TERM YIELD TRENDS

If we are interested in the potential for progress, we should look at some long time yield trends of major U. S. crops. These are for the period of 1933 to 1974.

Figure 1 shows that the yield of corn was fairly static until the early 1940's when hybrid corn was introduced on a commercial scale. Another major jump in yield was obtained in the early 1950's with the availability of relatively cheap fertilizer, especially nitrogen. An additional yield boost came with improved pesticides and improved mechanization in the late 1950's and early 1960's.

Figure 2 shows that little improvement in the yield of soybeans occurred until world demand increased after World War II. Although yield increases have not been as dramatic as with corn, yield improvement research gives encouragement to a strong future for soybean production.

Figure 3 shows that steady improvement in the yield of wheat has occurred since the World War II period with emphasis on new types such as the short strawed varieties of the last decade and the increased use of fertilizer and pesticides.

All of these yield trend lines show continued improvement. The degree of slope varies but offers encouragement for future development as research breakthroughs occur.

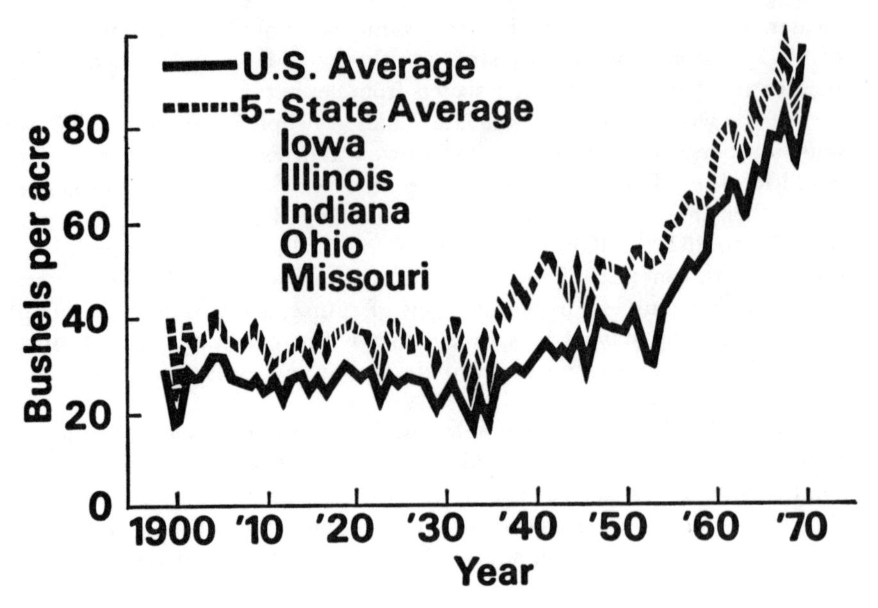

Fig. 1. Average corn yields by year.

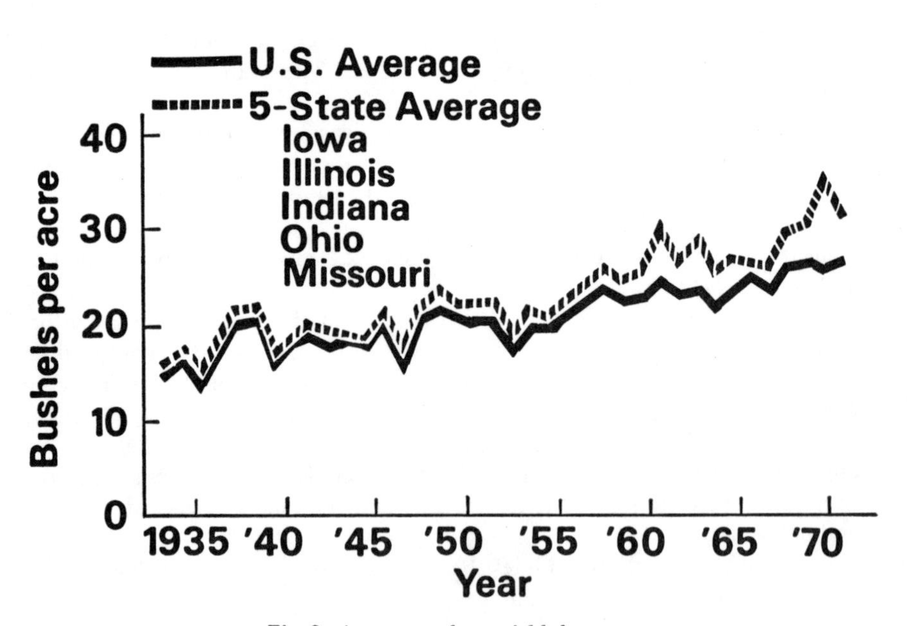

Fig. 2. Average soybean yields by year.

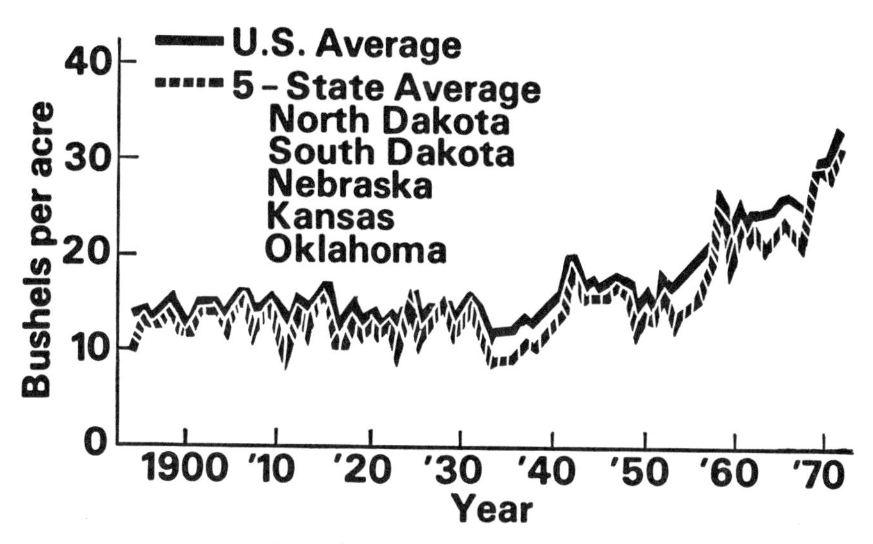

Fig. 3. Average wheat yields by year.

WHEAT IMPROVEMENT

Progress in the development of hard red spring wheat (*Triticum durum* L.) since 1926 to the present is shown in Table 1. 'Marquis,' a Canadian variety, was released in 1926. It was replaced by 'Thatcher' in 1935, 'Lee' in 1958, 'Chris' in 1964, and 'Era' in 1971; all four were Minnesota variety releases in cooperation with USDA. During the period 1926 to 1971, variety improvement led to a 79% yield improvement.

Table 2 illustrates the continuing progress being made in wheat breeding, with two experimental lines yet to be released showing a yield advantage over Era of 104% and 108% respectively.

We are often asked to justify the contribution of a research effort. The results in Table 3 show the contribution of Era wheat to the economy of Minnesota. A $300,000 investment has returned over $120 million dollars in four crop years, 1973 through 1976. Even with severe drought conditions in

Table 1. Spring wheat variety performance

Variety	Year released	Yield kg/ha*	% of Marquis
Marquis	1926	2,028	--
Thatcher	1935	2,230	10
Lee	1958	2,425	16
Chris	1967	2,735	35
Era	1971	3,623	79

* All tested in 1974 at 3 locations.

Table 2. Hard red spring wheat performance

	Yield		
Variety	kg/ha	Bu/acre	% of Era
Chris	2,681.3	39.9	80
Era	3,339.8	49.7	100
Exp. no. 1	3,474.2	51.7	104
Exp. no. 2	3,615.4	53.8	108

Table 3. Era (wheat) contributions to Minnesota agriculture

	1973	1974	1975
Acres of Era	971,300	1,625,000	2,025,000
Yield over other varieties	538 kg/ha (8 bu/acre)	336 kg/ha (5 bu/acre)	336 kg/ha (5 bu/acre)
Increased production	6,348,000	8,125,000	10,125,000
Price for 27 kg (1 bu)	$4.20	$5.00	$3.50
Increased income to growers	$26,587,680	$40,625,000	$35,437,500
Increased number of 0.454 kilogram (1 pound) loaves of bread	506,240,000	650,000,000	810,000,000

Table 4. Sources of wheat yield increase in Minnesota from 1940 to 1975

	% of Total Yield Increase
Breeding	
Yield	26-29
Disease resistance	19-22
Cultural practices	
Fertilizer, herbicides, etc.	19-26
Mechanization	
Equipment, timeliness of planting, harvesting	26-32

many parts of Minnesota this year, Era out-yielded other varieties by 201.6 to 336.0 kg/ha (3 to 5 bu/acre).

Critics of the "Green Revolution" point to failure of the "new seeds" because they needed the expensive technology of fertilizer, irrigation, and pesticides. Table 4 shows the relative contribution of the components of improvement. I believe this should dispel the expression of doubt, and offer encouragement to the LDC's. Dr. Heiner developed this analysis from information about yield improvement in hard red spring wheat. The results show that 26 to 29% of the yield increase resulted directly from breeding for yield, and 25 to 27% from breeding for disease resistance; i.e., breeding accounts for over half of the yield improvement. Improved cultural practices, use of fertilizer and pesticides, and irrigation, contribute 19 to 26% to improved wheat yields. Mechanization contributes 26 to 32%. Yet, mechanization is often the first thing thought necessary for improved food supplies in LDC's.

These figures point out that LDC's can use these "new seeds" to advantage, compared to the standard varieties now being grown, even though they might not have the latest technological advances at their disposal.

CORN IMPROVEMENT

Table 5 shows the importance of date of planting on the final yield. These results, obtained from Iowa studies, show that the yield advantage of early planting is one of the major factors in the purchase of larger equipment on Corn Belt farms. The timeliness of planting is important and it has strong income implications. Similar results have been obtained in long-term "cultural practice" studies at Minnesota.

Table 6 shows the yield of corn by hybrids of different eras grown from

Table 5. Effect of planting on corn yield in Iowa

Planting date	Flowering date	Yield kg/ha
May 1	July 23	8,173
May 10	July 25	8,047
May 20	July 31	7,544
May 30	August 5	6,978

Table 6. Yield of corn hybrids of different eras grown from 1971 to 1973

	Poor conditions		Good conditions	
Period hybrid developed	kg/ha	% increase over 1930	kg/ha	% increase over 1930
1930	3,709	--	6,538	--
1940	4,464	20	7,544	15
1950	4,778	29	7,670	17
1960	4,902	32	8,550	31
1970	5,972	61	8,990	38

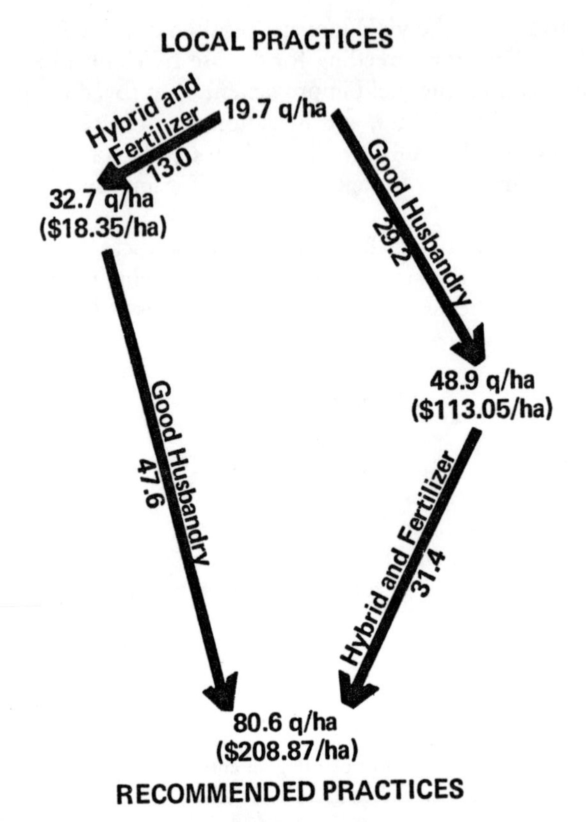

Fig. 4. Yield and income improvement from adaptation of production practices.

1971 to 1973 in Iowa under good and poor conditions (Russell, 1974). This again illustrates the "new seed" concept. On a percentage basis, the improvement potential under poor conditions was higher than under good conditions. The actual yields were all greater under good conditions.

Figure 4 illustrates what is possible in many LDC's by adaptation of recommended practices from other countries. Yields go through a series of improvements from the traditional peasant farming practices at 19.7 to 80.6 quintal/ha (17.6 to 71.9 hundred weight/acre) with all the possible technology being used. The value of the improved increase was $208.87/ha (Eberhart & Sprague, 1973).

Figure 5 illustrates the benefits from improved technology as a result of research and adoption of research results. Yields would have remained essentially unchanged during the period of 1945 through 1974 without this technology change.

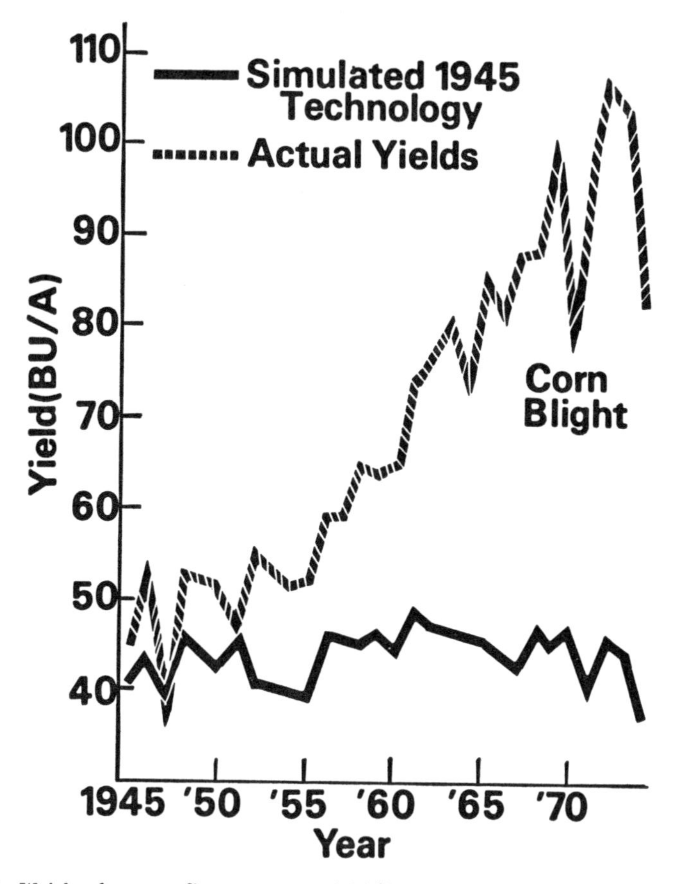

Fig. 5. Weighted average five-state corn yield (Ohio, Indiana, Illinois, Iowa, Missouri).

OTHER IMPROVEMENT RESEARCH

Additional examples can be shown by research under way at Minnesota. Wild rice research indicates that this wild species can be "tamed" and can result in new income from formerly abandoned land in long time financially depressed areas. Two new wild rice (*Oryza sativa* L.) varieties (Table 7), released in 1976 will provide about 660 to 880 kg/ha (588.7 to 785.0 lb/acre) more yield than standard varieties. Most importantly, they are 2 weeks earlier in maturity. Hopefully this will reduce bird and disease damage—a major threat to this new industry.

Continuing work on improvement of reed canarygrass (*Phalaris arundinacea* L.) makes this specie increasingly attractive as a forage for live-

Table 7. Wild rice yield test, Grand Rapids, Minn., 1975

Entry	Heading Date	Yield, kg/ha (lb/acre)
Johnson	8-5	821 (733)
M1	8-2	1,128 (1,007)
K2	8-1	983 (878)
Exp. 1	7-19	1,254 (1,120)

Table 8. New reed canarygrass performance

Kind	Percent Alkaloid	Steer Daily Gain per day	
	%	kg	lb
M-72	0.13	0.4	0.9
Rise	0.18	-0.6	-1.3

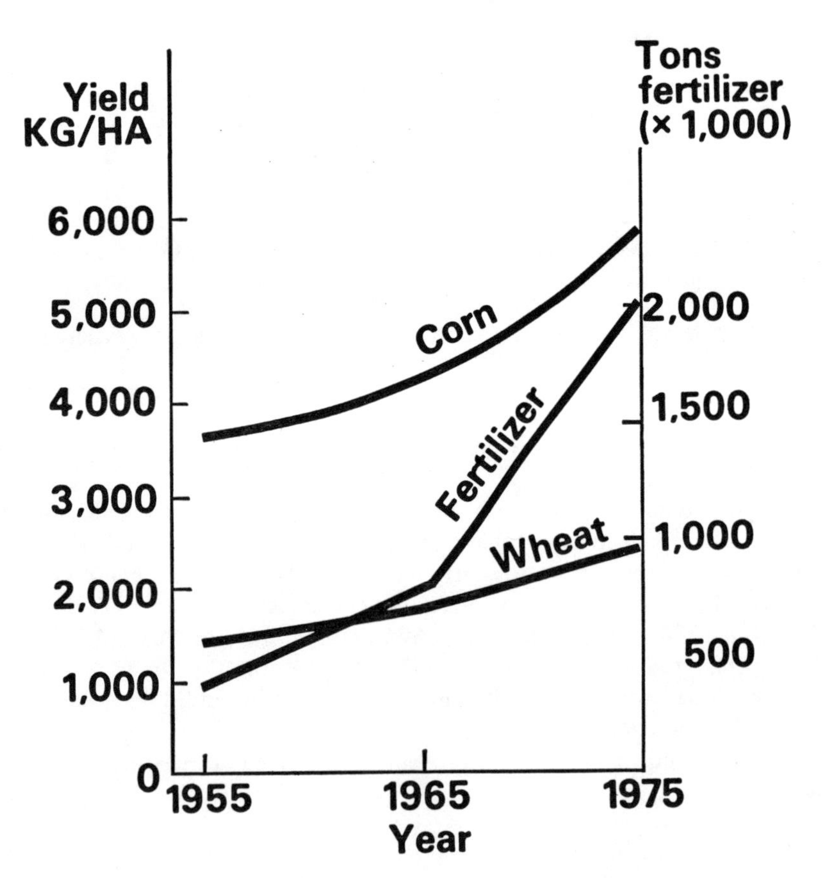

Fig. 6. Fertilizer use and corn and wheat yields.

stock. Table 8 indicates that the major improvement is from a lower alkaloid content resulting in a greater rate of consumption by animals, with average daily gain improved by 1.0 kg (2.2 lb).

The possible shortage of nitrogen fertilizer because of high prices for natural gas or limited availability bring greater need for research efforts in nitrogen fixation. Donald Barnes, Agricultural Research Service (ARS) alfalfa geneticist at the University of Minnesota, and his students have developed experimental strains of alfalfa (*Medicago sativa* L.) that in the greenhouse have twice the nitrogen fixing capacity as in available strains. As we look at a possible return to the greater use of rotations, this type of alfalfa can have a significant effect on succeeding crops of corn or small grains.

Figure 6 shows the correlation between fertilizer use and yields of corn and wheat in Minnesota from 1955 to 1975. These increased yields are no doubt a combination of factors including improved varieties, management practices, and pesticides as well as increased fertilizer use. However, the results do illustrate the role of improved plant nutrition on grain yields with the effect being greater in the case of corn than of wheat.

CONCLUSION

These illustrations of long-term continuing research programs in yield improvement offer encouragement to meet the challenge of a hungry world. Our base is solid and our past contributions deserve great plaudits, but what we do in the future will determine the destiny of mankind.

American and world agriculture has obtained its productive capacity in part from the varied programs of classroom teaching, research, and extension. For nearly 70 years, the profession of agronomy (crop and soil science) has made major contributions to developing the food and fiber system we have in the United States and the developed countries of the world. Major contributions have been improved plant varieties in all major field crops and understanding the way plants grow and respond to different cultural practices, such as use of fertilizers, row spacing, pesticides, and tillage. The basic understanding of plant processes such as photosynthesis, genetic control of disease, and insect resistance have been essential components. These sciences, coupled with other disciplines so vital to modern agriculture, make it possible for 800,000 farms in the United States to provide 88% of U. S. food and fiber supplies, and 25% of world supplies. Such progress offers challenge, opportunity, and hope to the developing nations of the world.

LITERATURE CITED

Eberhart, S. A., and G. F. Sprague. 1973. A major cereals project to improve maize, sorghum, and millet production in Africa. Agron. J. 65(3):365-73.

Russel, W. A. 1974. Comparative performance for maize hybrids representing different eras of maize breeding. p. 81-101. 29th Annual Corn and Sorghum Research Conference, American Seed Trade Association, Washington, D. C.

Crop Physiology to the Year 2000

R. W. Howell

We are in our Nation's third century. In a very short time it will be the year 2000–23 years away. There are few readers of this paper whose present age will have doubled by 2000.

It is not very long. And if anything can be predicted with confidence it is that the future will be haunted by the need for food. Can the world feed its people? The question is political, social, religious–and especially agronomic. Agronomists must make it possible to increase the production of food or the political, social, and religious questions will be moot. Physiologists must play a crucial role.

Modern crop science rests on the discoveries of crop physiologists, sometimes calling themselves plant physiologists, botanists, or agronomists and they date back a century or more. Knowledge of mineral nutrition and soil fertility is derived from work of J. Liebig in the 19th century. The plant nutrition studies of D. R. Hoagland and his successors, D. I. Arnon and P. R. Stout, and of H. Lundegardh were basic to modern agronomic practice. The movement of materials in plants was described by Stephen Hales. Photoperiodism was recognized by W. W. Garner and H. A. Allard before 1920, and was later described by H. A. Borthwick, M. W. Parker, S. B. Hendricks, and others. Hormone action was described by F. W. Went and others quite a long time ago.

After World War II plant physiology entered a quantitative phase. Before that it had been mainly descriptive. Postwar plant physiologists concentrated on mitochondria, membranes, and other organelles, and on enzymes, tissues, and differentiation. The universality of biochemical principles became more certain. For many reasons, including the precision and rapidity of biochemical experiments and the availability of money from nonagricultural sources, plant physiologists were not very interested in the physiology of crop production. "Whole plant" physiology declined in popularity.

R. W. Howell is head of the Department of Agronomy at the University of Illinois, Urbana, Ill.

It was hard for agronomists to see the relevance of much of this work to their problems. So, agronomy departments began to build their own physiology units. There were great expectations. Crop physiologists were challenged to find the factors that limit yield and to discover what would be necessary to raise yield potentials. How to make the "breakthrough" was a frequent question. It seemed that any unsolved problem should be attributed to "physiology." Somehow, that seemed informative and reassuring —about like the physician saying "It's a virus that's going around."

Probably the most successful of crop physiologists have been the weed scientists. They have been so successful that not one, but two new societies have appeared: The Weed Science Society of America and the newer Plant Growth Regulator Working Group, which is concerned with chemical regulation of growth in a sense much broader than weed control. In those areas there were simple tests of treatment effectiveness. The weed died, the fruit fell, or the flower bloomed in unequivocal response to treatment.

But the plant system is complex and usually simple tests of treatment effectiveness are not available. Unlike microbes, plants have a multicellular structure differentiated in form and function. Effects may be expressed at a site other than that of treatment, and may be long delayed. Unlike animals, plants are exposed to the environment with little ability to reduce exposure. The relationship of plants to the soil is unique to the plant kingdom.

This brings us to the question, "What should crop physiologists do in the future?" The rest of this century will be a period of heightened expectations and diminishing resources. The food production system at home and abroad will be challenged to greater productivity in the face of declining supplies of fossil-related fuel and equipment. The crop physiologist will share the responsibility of finding ways to make crops, that is, communities of plants, produce more per unit of land.

The challenge is very clear. The physiologist must identify problems more specifically; he must find general problems that are important, and components of those problems that are physiological. This is the first task—to identify, i.e. define, specific physiological problems that are important.

Choosing your problem is a very personal responsibility. If you choose wisely, you have a good chance for success. If unwisely, you can wander around endlessly in the backwaters of science. It is not necessary to make a career of a thesis problem nor to strive indefinitely in a project that is unproductive. You of course must have some concern about finances, but do not prostitute yourself. Not everyone should march to the loudest drum. Some research demands large groups and expensive equipment; some does not. Choose your own area as wisely as you can.

Crop physiologists should be practical, but research on basic biological functions in crop plants can add *important* (it may even be "useful") knowledge. Crop physiology needs more study of the principles of growth of crop plants, more attention to the biochemical as contrasted with the external or environmental. The latter are very important, and the lack of external or environmental plant physiology is what got agronomy departments into physiology. We need more study of basic metabolism by physiologists with an agronomic orientation. R. H. Hageman and his students and former students have explored nitrate metabolism extensively. The classical discovery by O. E. Nelson and E. T. Mertz of the protein variance associated with opaque-2 gene in corn (*Zea mays* L.), the discovery by B. R. Stefansson of rapeseed without erucic acid and the discovery of high oleic acid safflower (*Catharmus tinctorius* L.) types by P. F. Knowles show what is possible when principles of physiology are understood. Use of basic knowledge has so often followed discovery that to dwell on the point is to belabor it.

The yield of an economic product makes a plant a crop. Flowering, fruit and seed set, translocation, partitioning of photosynthate, and synthetic activities and energy relations in storage organs still challenge the physiologist. It is a vast area, and its importance justifies crop physiology as a branch of agronomy.

It seems quite clear that plants have endogenous systems to ensure that some normal seeds are produced. If conditions are favorable, many normal seeds are produced; if unfavorable, fewer are produced. That has been known for a long time, but I think not much is yet known about how the control signals originate and how they work. Useful concepts, such as "sinks", do not so much reflect knowledge as identify unknowns.

Yield physiology in all its aspects must concern all agronomists. Yield is what the farmer has to sell.

Closely related to yield is quality. I have cited examples of discoveries related to seed quality in corn, rapeseed, and safflower. Forage physiologists are much concerned with quality also, as they should be. The efficiency of animal feed, especially poultry, has been improved through better grain mixtures. For grazing animals, improvement of forage or pasture quality is needed.

Consideration of crop quality has not yet counted for much in agricultural markets. But our concern is with the future. And quality, especially nutritive quality, will be important.

Life is dependent on biological nitrogen fixation, mostly symbiotic. There was little urgency for research on nitrogen fixation when yields were low, commercial nitrogen was cheap, and nitrogen supplies seemed inexhaustible. Commercial nitrogen is not cheap now and the raw materials to make it are certainly not inexhaustible. Nitrogen fixation systems that are more efficient in the presence of soil nitrogen need to be found. Nitrogen fixation by non-legumes has been reported. This area should be explored vigorously.

Another vital subject is soil/plant interactions. The soil chemist speaks of cation exchange and mass flow, of the "fixing" of cations so that they are unavailable. The physiologist often works in simple nutrient solutions. What goes on at the root/soil interface? Why do some plants prefer acid soils and others alkaline? What energy relationships permit selective ion uptake by roots seemingly against concentration gradients? Do microorganisms, such as mycorrhizal fungi, play a significant role in the release and uptake of nutrient elements? J. Brown, C. Foy and their associates at Beltsville, Md. have done elegant studies of the difference between iron-efficient and iron-inefficient varieties of crop plants and have shown the potential for breeding for resistance to acid soils. E. Epstein has found cereals that thrive in salt solutions. Such studies should be done with other crops.

Concern with environmental quality adds a dimension to mineral nutrition and fertility. The potential for pollution from plant nutrients, the cost of nutrients in energy and dollars, and the finite supply of nutrient resources add urgency to plant nutrition studies. The need to reclaim disturbed lands and the problem of what to do with mountains of waste introduce a host of possible troubles with toxic elements such as cadmium, selenium, lead, and mercury, as land application appears to be the waste disposal option of last resort.

Water is the main factor limiting crop production—sometimes too much, more often too little. Even rice (*Oryza sativa* L.), the mention of which evokes visions of paddies, is a crop in which drought stress is significant. Water, being mined in areas such as the Texas Highplains, is limited by the power costs of irrigation in Nebraska, and is simply unavailable in some places. Possible adverse changes in climate pose a threat of unknown seriousness. Can we learn how to use water more effectively? *That* is a high priority question.

Crops of the future will have to perform well under stress. I have mentioned water stresses. There are also temperature stresses—either too much heat, or not enough. Heat stress can sometimes be lessened if there is enough water. Cold stress calls for protection ("smudge pots" under citrus trees or snow on small grains). Perennial trees and winter crops must be cold tolerant. Chemical modification of membranes may increase cold tolerance and should inspire additional study of biochemical mechanisms of stress tolerance.

There are stresses due to disease, pesticides, wind, and hail. We rarely know as much as we would like to about minimizing the effects of stress. It is hard to isolate stress physiology from other aspects of crop production, including genetics. But it remains an urgent need to find out more about how plants can cope with stresses.

Photosynthesis is the most fundamental of biological processes, typical of green plants and almost unique to them. It is the pump that makes life possible. Everything else is downhill energetically. Yet the great discoveries of photosynthesis by R. Hill, O. Warburg, R. Emerson, C. Van Niel, M. Calvin, and others have not shown how to manage photosynthesis for crop production. It is not without reason that much attention in the last decade has been directed toward better descriptions of the nature of photosynthesis in plant communities, of the reasons why some species are more efficient than others, and of ways to improve that efficiency. The C-4 pathway of photosynthesis and the significance of photorespiration were discovered largely because agronomists sought to manage photosynthesis for increased crop productivity.

So far, agronomic approaches to photosynthesis have not answered all the questions. Intraspecific differences in photosynthetic rate which could be exploited by plant breeders have not been found. But the lack of quick progress does not mean that photosynthesis is unimportant. Increased research on photosynthesis continues to be recommended in many reviews and reports, emanating not just from agriculture, but also from those probing production of "biomass" as an energy resource. Some of this photosynthesis research needs to be done by crop physiologists.

And then there is the whole area of abnormal physiology, usually referred to as pathology. The line between physiology and pathology is obscure. Seed quality is influenced by environmental and pathogenic factors. The susceptibility of T cytoplasm of corn to the southern corn leaf blight is associated with sensitivity of mitochondria to a toxin produced by race T of *Helmenthosporium maydis.* Brown stem rot in soybean (*Glycine max* L. Merr.) produces a substance that almost instantly reduces vascular conductivity. Recently plant physiologists in Colorado and Australia have identified the active materials in the resistance of soybeans to phytophthora rot.

There are no doubt many other examples. A comparative study of normal and abnormal physiology can be very informative as to normal physiology.

I have reviewed a lot of conventional physiology research areas. They continue to be important because crops still require light, water, carbon dioxide (CO_2), oxygen (O_2), mineral nutrients, and an environment within narrow limits. The unconventional approach also deserves consideration, tissue and cell culture for example.

Regeneration in plant cells and tissues is more common than in animal cells, and less controversial. It was more than 25 years ago that F. C. Steward of Cornell University successfully regenerated a carrot plant from a single cell isolated from the root, and nearly as long since Carlos Miller discovered cytokinins with tobacco tissue cultures at the University of Wisconsin. Cell and tissue culture techniques have been powerful tools in the study of differentiation and now are used to modify amino acid composition with potentially great improvement in food value of plants. Culture of organelles such as chloroplasts may accelerate the understanding of photosynthesis.

Plasmid transfer of genetic information has been accomplished in microorganisms. Nitrogen fixation may be amenable to unconventional transfer of genetic information. Disease resistance, or modification of oil or protein might be achieved through a genetic bridge, such as plasmid transfer. Who will be wise, lucky, and bold enough to create a C-4 photosynthesis in wheat or cotton or soybeans, or nitrogen fixation in cereals by plasmid transfer or some other technique not yet imagined?

These are enough of examples of research. They could be stated differently and there could be others.

Physiologists in the universities also have teaching responsibility. Learned papers have their place, but so has teaching. Any agronomist needs to know something of the principles of physiology—why plants and plant communities behave as they do. The crop physiologist has a key role in linking advances in physiology and biochemistry to the problems of crop production and in transmitting this knowledge to new generations of students. Physiology curricula should not be designed solely for physiology majors.

Agronomy departments have physiologists and recognize crop physiology as a subdiscipline of agronomy because there are problems that require the skills and techniques of the physiologist to solve. Physiologists should be alert to cooperate with colleagues in other disciplines. Disciplines are creations for human convenience; problems do not always come neatly packaged in a single discipline. If a technique useful to a colleague is discovered, for example a physiological trait which the breeder can manipulate, so much to the good. But that is not all there is to collaboration and cooperation, nor does cooperation consist of conducting experiments and reporting the results with the conclusion that the information should be useful to the breeder. "Biochemical genetics" is a common phrase, but an uncommon practice. Genetic variability in physiological traits has been shown in many studies. Great progress in biochemical genetics demands collaboration of physiologists and breeders.

Physiologists should get to know and become involved with breeders, soil chemists, microbiologists, weed scientists, extension specialists, pathologists, and others. An extension specialist may not often seek out a physiologist for advice, but he is more likely to do so if he knows what the physiologist is doing.

Physiologists need to do all this while continuing professional association with other physiologists. There are about 450 members of agronomy, crops, or soils departments who are members of the American Society of Plant Physiologists, and many more who are members of the Weed Science Society, or the Plant Growth Regulator Working Group. If you neglect these professional groups, your physiology will become stale.

Do not look too much to your administrators to tell you what to do, nor to some committee, panel, or board. Administrators will have done well if they recognize broad areas. The buildup in crop physiology in agronomy departments over the last 10 to 15 years resulted from administrative decisions, sometimes urged by other agronomists, and were implemented with funds from industry or appropriations that followed recommendations from administrators, scientists in other disciplines, and industry spokesmen. That is about the most to be expected from the administration. And it is quite a lot.

Brilliant research comes from scientists who have thoroughly reviewed what is known, carefully developed hypotheses as to important questions with unknown answers, and rigorously tested the hypotheses. If your work is good, it will be recognized; if not, you will probably be found out. You should be your own severest critic.

I recently attended a meeting where a group of panelists lamented their inability to make their directors appreciate the importance of their work. By the end of the session, they had convinced me that the directors were right.

Crop physiology research should develop an improved understanding of how crop plants grow. With some outstanding exceptions, crop physiology has not been very biochemical. It should become more so. It should answer the following questions.

How do crop plants transform the raw materials of water, carbon dioxide, and inorganic nutrients into protein, carbohydrates, and lipids? How do they concentrate metabolic products in specific tissues? How do they adjust to environmental variables of light, temperature, water, and wind? What are their defenses against disease, insects, weeds predators, and competition among themselves? All of these unknowns, and more, challenge the crop physiologist in Century III.

The Role of Agronomists in International Agricultural Development

N. C. Brady

In the early 1970's, the United States, for the first time in its history, was affected directly by world food shortages. In previous decades, the problem of food surpluses, not food deficits, had plagued America. The average American generally learned of the world food problems by reading newspapers and popular magazines that presented stories and photographs of the very young and the very old, dying from hunger or malnutrition in densely populated, low-income developing nations of the world. American efforts to provide food aid were sympathetic and generous, but thc problem seemed remote.

Then, abruptly in the early 1970's, the cornucopia was threatened. The severe droughts in Asia, the Soviet Union, and parts of Africa drastically limited food production, and brought the unsettling awareness that the U. S. food reserves, which had previously been a political headache, had reached unprecedented low levels.

The grain and other food reserves, which had provided an effective buffer in previous food crises, were inadequate to prevent widespread food shortages and the concomitant spiraling of food prices. Now, both the poor countries of the world and the affluent nations of North America and Europe were similarly affected. The harsh reality of the world food-population crisis had reached the American neighborhood. The well-stocked shelves of supermarkets were no longer bountiful, and housewives protested vigorously against skyrocketing food prices. The nation realized that the food-population problem was international and that America had a responsibility to help solve it.

A positive effect of the world food shortages of the early 1970's was the sharp change in the attitude of the United States towards the food supply problem. Previously, most political leaders and the general public had little concern about that international problem. Now, the situation has changed.

N. C. Brady is Director General of The International Rice Research Institute, Los Baños, Laguna, Philippines.

Following the lead of some thoughtful statesmen and scientists who had long placed high priority on the world's capacity to feed itself, increasing numbers of educators, businessmen, and political leaders have shown support for greater U. S. participation in efforts to increase world food supplies and to stabilize population increases.

Evidence of this increased support was seen in the keynote address by Clifton R. Wharton, president of Michigan State University, at the 1976 annual meeting of the American Society of Agronomy (ASA) (see Chapter 1). He discussed the topic of the world food supply and the role of agronomists and other scientists in helping to produce more food. He and other agricultural leaders are concerned with the implementation of a historic piece of American legislation, popularly known as Title 12. Title 12 provides the mechanism to utilize the resources and scientific talents of U. S. universities, other agricultural research organizations, and American industry and business to help improve world food production.

The prestigious National Academy of Sciences/National Research Council's Board on Agricultural and Renewable Resources (BARR), in response to a request from the office of the U. S. President, has identified changes that could and should be made in the focus of U. S. agricultural science to help solve the world's food problems (Board on Agriculture and Renewable Resources, 1975). President Ford requested an even more extensive study, which is now under way at the U. S. National Academy of Sciences.

The American Society of Agronomy was among the first of the scientific societies to focus its attention on world food problems. In fact, "Food for Peace" was the theme of the first of the Society's special publication series, initiated in April 1963. I had the privilege of serving as co-editor of that publication, which featured papers by scientists, research administrators, and political leaders concerned with this vital topic.

There are good reasons why we soil and crop scientists have taken the leadership in drawing the attention of the world to the international problem of food production. Our job is to help farmers produce as efficiently and economically as possible. The record will show that we have not been idle in our attempts to do so.

But if we are to help farmers in the developing nations increase their capacity to produce food, we must focus our attention on those areas of the world where food shortages are most severe. It means that priority must be given to the tropics where most of the food-deficit countries are located. Special emphasis must be placed on Asia, where three-quarters of the world's hungry live (Table 1), and on Latin America and Africa, where vast areas of unused or underused soil resources still exist.

Some notable achievements of agronomists exemplify what has been done and point to future tasks if we are to help increase food production internationally. Let us consider first the accomplishments of plant breeders and the opportunities they now have to develop new and improved food crop varieties.

The development and spread in the early 1960's of the new high-

Table 1. The Food and Agricultural Organization (FAO) estimates of the number of people who had insufficient protein-energy supply in 1970, by regions. (Data presented at the United Nations World Food Conference, quoted by Poleman, 1975)

Region	Population (million)	Below lower limit	
		%	No. (million)
Developed	1074	3	28
Developing†	1751	25	434
Latin America	283	13	36
Far East†	1020	30	301
Near East	171	18	30
Africa	273	25	67
World	2825	16	462

† Excludes Asian centrally planned economies.

yielding varieties of rice (*Oryza sativa* L.) and wheat (*Triticum aestivum* L.) drew the world's attention to what agronomists and other production-oriented scientists can do to help alleviate world food shortages. These new cereal crop varieties had the genetic potential to far outyield their traditional counterparts. They responded dramatically to management and chemical inputs. They raised the sights of even the humblest of farmers. They set new goals for agronomists and other scientists working to help farmers in all parts of the world. Likewise, they convinced decision makers and program planners in both developing and more advanced countries that science was an essential and indispensable ingredient in the blend of components needed to increase food production.

Two research efforts illustrate the extent of the success of modern cereal varieties. The first is an analysis by Evenson (1974) of the production increases and economic values attributable to the new rice and wheat varieties. Even with the relatively low food prices prevalent in 1972, Evenson estimated the *annual* value of the increased production that can be attributed to the new rice and wheat varieties to be 2.9 and 1.1 billion dollars, respectively (Table 2). The significance of that finding explains why regional and

Table 2. Increase in production associated with the use of high-yielding rice and wheat varieties, Asia and the Mid-East (Evenson, 1974)

Crop year	Increase in			
	Production (%)		Value (million US$)	
	Wheat	Rice	Wheat	Rice
1966/67	1.5	1.0	58	148
1968/69	18.3	5.5	732	784
1970/71	22.1	12.7	884	1798
1972/73	28.2	20.7	1128	2933

international banking institutions view research inputs as good investments in economic development.

The remarkable increase in rice yields and production in Colombia that resulted from the development and adoption of new rice varieties is another striking example of the success of the research efforts of plant breeders. Since 1967, when the rice variety improvement and introduction program was intensified in Colombia, rice yields have increased nationwide from less than 3 metric tons/ha (1.3 tons/acre) to more than 5.4 metric tons/ha (2.4 tons/acre). Jennings (1976) reports that "in Colombia alone and in one year alone, 1974, the added production resulting from the introduction of new technology was valued at US$230 million. . ." Inasmuch as rice research expenditures in Colombia were less than $1 million annually, Jennings' observation, ". . .that is a splendid return on any investment," is indisputable (Fig. 1).

Plant breeders face even more significant horizons ahead. They are now asked to incorporate increased insect and disease resistance into modern varieties of cereal crops. While this is a task to which the plant breeder is accustomed, it carries a formidable responsibility, particularly for the low-income countries of the tropics. In these countries, the availability of pest

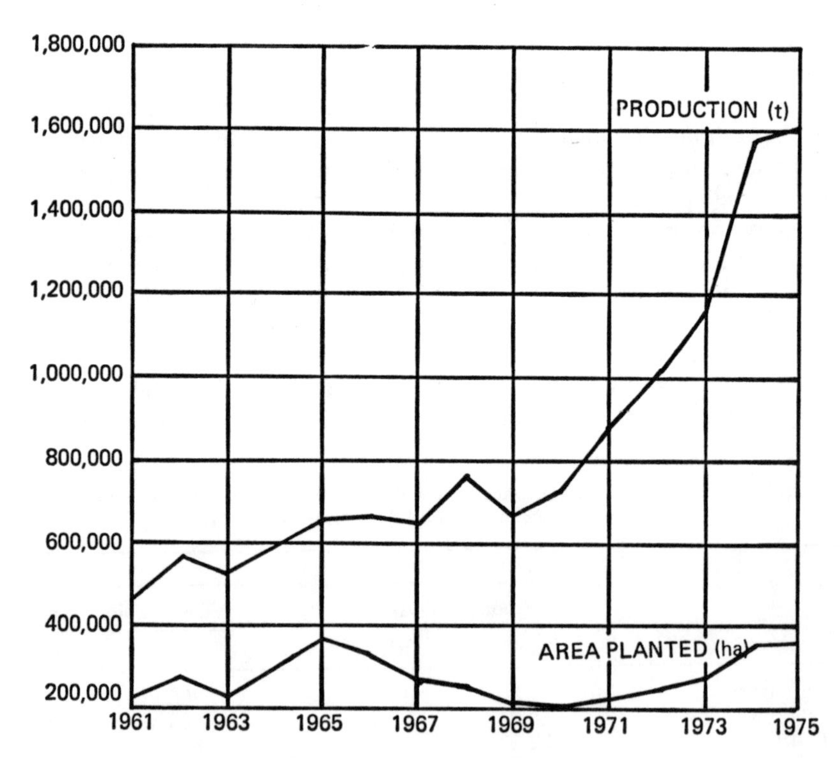

Fig. 1. Rice production in Columbia rose sharply after the introduction of improved varieties in the late 1960's (Jennings, 1976).

management methods other than through host resistance is grossly inadequate. Pesticides are expensive and farmers often lack the expertise and technical knowledge needed to use them effectively. Furthermore, the daily rains in the humid tropics wash away the chemicals before they can be effective. The condition necessitates the use of economically unsound quantities to effectively control insects and diseases. Thus, agricultural production specialists and farmers turn to plant breeders for assistance.

The demand for farming systems with greatly increased cropping intensities provides plant breeders further opportunities and responsibilities. Improved cropping systems depend on the availability of varieties with good yield potential and short growth durations. Plant breeders have developed, for the tropics, high yielding rice varieties that require only 70 to 80 days from transplanting to harvest, and only about 100 to 110 days from basic seeding to harvest. The latter figure contrasts with the 150 to 170 days of growth duration common in the old traditional varieties. Thus, commendable progress has only begun. With these few rice varieties as prototypes, a series of new short-seasoned varieties must be developed for the diverse environments of the tropics.

The developed nations have financial, biological, and chemical resources to modify the environment so that good crops can be grown. They require water, fertilizer, lime, and other amendments to alleviate the soil and climate deficiencies prevalent in a cropping area. In some cases, they have converted infertile and unproductive soils into productive soils.

In most developing countries, however, it is generally uneconomical to modify the soil environment to suit the requirements of existing crop varieties. Thus, plants must be tailored to fit the environment, and not the reverse. Conseauently, the plant breeder is called upon to develop varieties that can resist or tolerate the adverse soil and climate environments where crop plants are to be grown.

The manipulation of rice germ plasm provides an example of the important role that plant breeders can and must play in international agricultural efforts to improve food production. Rice varieties exist that will grow well on soils with pH values of 4.0 or less; others are known to tolerate soil pH values of 8 or higher; still other rices will grow in areas that have soils with high salt contents.

Similarly, rice varieties are available which produce well under the conditions of cold air and water existing in northern Japan. Still other rices thrive under the hot dry climates of Pakistan and Iran (Chang & Vergara, 1972). Genetic resources of rice also include varieties that tolerate or resist low soil moisture conditions and drought. Additionally, rices that can withstand complete submergence for short periods have been and are being developed; they are critical in areas where unexpected floodwaters can destroy crops.

The availability of those rice genetic resources with desirable traits indicates the possibility of developing varieties suited to specific ecological conditions where rice is grown. The job of producing such varieties is complex and requires the help of problem-area scientists such as entomologists, plant pathologists, soil scientists, and plant physiologists. At the International Rice Research Institute (IRRI) these scientists do 70% of the genetic evaluation and utilization. But the plant breeder is the nucleus of the team.

The nutritional quality of new food crop varieties is important and will be a high priority area for plant breeders in the future. Reaching the goal of elevating the quality of plant protein will require concerted efforts. In areas where they do not compete with man for edible crop plants, animals are used to convert to high quality animal protein forage and other grasses inedible to humans. In areas of high population density, however, plants will continue to furnish the preponderance of food in the future, and plant breeders bear the responsibility of developing crop varieties with high nutritive quality.

Plant breeders and soil and crop management researchers also need to expand their research on grain legume crops. The cereal grains, especially wheat, rice, corn (*Zea mays* L.), and sorghum [*Sorghum bicolor* (L.) Moench.], have received priority in both the developing and the more developed countries. While this emphasis may well have been justified for economic reasons, current and future nitrogen requirements in food production as well as the need for food protein in low-income countries justify that higher priority be given to the protein-rich grain legumes.

Despite the recent progress, modern crop varieties still lack adequate genetic diversity. Dramatic evidence of the widespread genetic uniformity is the corn leaf blight epidemic in the United States in 1970. A subsequent review by a committee of scientists convened by the Agricultural Board of the National Academy of Sciences/National Research Council revealed a genetic vulnerability of most major crops on a worldwide basis (Agricultural Board National Research Council, 1972). Modern varieties of major crops were found to have considerable genetic uniformity, which makes them vulnerable to epidemics of pests and diseases. This is especially critical in developing countries because a few recently developed high-yielding varieties with relatively narrow genetic bases are replacing large numbers of traditional varieties. The new varieties respond to heavy fertilizer and water management practices that result in luxuriant plant growth and concomitant vulnerability to insects and diseases, especially those which attack crop foliage.

A perilously narrow genetic base is found for most modern semidwarf rice varieties. For example, a recent survey of plant breeders in India shows that 94% of the local semidwarf modern varieties are progenies of IR8 or TN1 (International Rice Research Institute, 1975a). Plant breeders must take steps to broaden the genetic base of resistance or tolerance to pests, diseases, and other factors which handicap food crops. The world's genetic resources of these food crops must be conserved and systems developed to disseminate such resources to interested scientists in both developing and more developed countries.

In the next decade, the plant breeder must seek new and innovative methods that can markedly shorten the time required to develop and test lines in varietal improvement programs. The BARR report (Board on Agriculture and Renewable Resources, 1975) recommends the development of "techniques for genetic manipulation beyond those of conventional plant breeding, including in vitro techniques for asexual approaches, and broad crosses between crop species." Single cell culture, anther and pollen culture, and somatic hybridization techniques must be thoroughly explored. While some success with these techniques has already been achieved with a few plants, further research is needed to fully explore the possibilities those techniques present (Carlson & Palacco, 1975). Plant breeders will need the assistance of geneticists, biochemists, and plant physiologists to develop these innovative methods.

Turning now to the role of the soil scientist, accomplishments of the past and opportunities for future inputs are likewise noteworthy. The development of many technological packages that helped farmers realize the genetic potential of new crop varieties rested with the soil scientists, who demonstrated the responsiveness of the new varieties to chemical fertilizers, particularly nitrogen. Soil scientists have also begun to identify and to better characterize the soil conditions necessary for good yields of food crops in the tropics.

In no other research area can soil scientists make a greater contribution than in the conservation, management, and efficient utilization of water. It has been aptly stated (Michigan-Kettering, 1975) that "water, in shortage or excess, is the most common limiting factor in production." Research and technology transfer are essential to increase water use efficiency, whether in developing countries where grossly inefficient irrigation systems are common or in the much more vast dryland areas and in rainfed humid and semihumid regions.

Crop and water management systems must be developed and put into practice to check soil erosion on the millions of hectares of rolling and steep lands in the humid tropics where population pressures have forced intensive cultivation. Attention must also be given to drainage and irrigation practices for areas of excess salinity, to prevent further salt accumulation. While some further research may be needed, marked improvements in water management can be obtained by merely putting to work what we now know.

The rising costs of fossil fuels have reemphasized the need for soil scientists to improve the efficiency of fertilizer utilization by crop plants. The type of work to be done is exemplified by the research conducted by soil scientists at IRRI on the efficiency of the use of nitrogen fertilizers in rice paddies. Drawing on the earlier work of scientists in India and Japan, IRRI scientists have shown that placement of fertilizers in concentrated bands or pellets in the reduced zone of the soil in the rice paddy doubled the efficiency of the rice plant in using nitrogen. Yields realized with 50 kg N/ha (44.6 lb N/acre) applied by the subsurface localized method in the root zone are similar to yields achieved when 100 kg N/ha (89.2 lb N/acre) is broadcast in the traditional manner (Table 3).

Soil and crop scientists must now expand their efforts to determine the fate of fertilizer chemicals added to soils of the tropics. It has long been

Table 3. Higher increases in yield were obtained with 60 kg N/ha concentrated in mudballs and applied to the rice root zone, than with 100 kg N/ha applied by the topdressing method used by farmers. IRRI, 1974 dry season (Data from S. K. De Datta quoted in International Rice Research Institute, 1976b)

Fertilizer rate (kg N/ha)	Yield (metric tons/ha) with N applied as		Efficiency of N (kg rice/kg N)	
	Mudball	Topdressing	Mudball	Topdressing
60	8.0	5.8	53	23
100	8.4	6.6	38	21

known that a major portion of the fertilizer nitrogen applied to soils cannot be accounted for by crop uptake, soil accumulation, or leaching losses. Soil scientists must seek a better understanding of the processes by which these losses occur and develop practical management practices to minimize them.

The biological fixation of nitrogen is another area where knowledge is needed. Research to gain a better understanding of its mechanism and the conditions under which the process occurs must be undertaken. Ways to increase its rate must be found. Concerted efforts will be needed to ascertain the full significance of nitrogen fixation in the rhizosphere with nonlegumes, and the role of algae in the biological fixation of nitrogen under the hot humid conditions in some tropical areas. The soil scientist will have a key role in all those investigations.

The need for soil scientists to work in international agriculture has become more evident with the spread of the new improved crop varieties. All too often, the performance of the new varieties is only slightly better than that of traditional varieties. Soil scientists discovered that the negligible response of both new and traditional rice varieties to added fertilizers in some locations in Asia was caused by abnormal soil conditions. They also found that iron toxicity limited rice production in areas of Sri Lanka, India, and Vietnam.

Soil scientists have discovered that zinc deficiency is a major problem in rice culture in some parts of the Philippines and India. Researchers found that in areas with zinc-deficient soils, dipping the rice seedlings in a 5% solution of zinc oxide before transplanting increased the yields on fertilized plots 8 to 10 fold or from 0.5 metric tons/ha (0.2 ton/acre) to 4 to 5 metric tons/ha (1.8 to 2.2 lb/acre) (International Rice Research Institute, 1973). Scientists in India, Indonesia, and other Asian countries have identified large areas of neutral or slightly alkaline soils where zinc deficiency is prevalent. Such problem soil areas must be identified and means to overcome the crop limitation they impose need to be developed.

Much work needs to be done in tropical regions of the world to more properly characterize soils for:

1) capacity to supply nutrients,
2) capacity to fix added nutrients,

3) oxidation-reduction status, and
4) ability to supply available soil moisture to crop plants.

Soil surveys and land-use capability inventories are woefully inadequate in most areas of the tropics. Both scientists and national agricultural production planners must know the types of soils at their disposal because various factors often limit the rate at which modification of existing unproductive soils can take place. Information on soil types is critical in increasing cropping intensities in the tropics, especially in combinations of paddy or flooded rice and upland crops. Alternate oxidized and reduced soil conditions accentuate chemical toxicities and deficiencies in some kinds of soils. These soil areas should be identified and mapped to assist national agricultural production planners.

Soil surveys are needed to characterize not only the areas currently under cultivation but also the vast uncultivated areas of the tropics that have potential as cropland. The accurate assessment of the world's maximum food production potential depends in part on the characterization of its soils.

Agronomists have a key role in the development of improved methods of weed control. Weeds are a serious problem in the humid tropics where the high total rainfall and the frequency of rains, often coupled with crude tillage and cultivation implements, make weed control difficult. Furthermore, the short-statured wheats and rices do not provide strong competition for weeds, a fact that compounds the problem of weed control. Many farmers with relatively small holdings do not have access to modern methods of tillage and cultivation that characterize farming in temperate zones. Consequently, weed control generally must be done manually with the use of small hand tools.

In the tropics, agronomists use herbicides with a degree of success. However, the cost of herbicides and the limited availability of expertise needed to assure their proper and timely application preclude their widescale use in the tropics. Nevertheless, the development of chemical and other means of weed control is essential to realize the productive potential of the new crop varieties being developed in the research programs of national and international agricultural research centers.

Manipulating the cropping patterns in relation to the weeds to be controlled is an effective and practical method of weed management. For example, including corn in a pattern which in the past involved only lowland rice shifted the weed population from difficult-to-control perennials to more easily controlled annual weeds (Table 4).

Agronomists also have important roles to play in the development of improved systems for farmers in tropical regions, where the potential to double and even triple crop production has yet to be exploited. The absence or inadequacy of implements to quickly prepare the soil and seed bed, coupled with the use of varieties which require long growth periods, has prevented the efficient use of the abundant water and sunlight available in the tropics for crop production. Only one crop per year is common even in areas where the climate is favorable for crop growth up to 250 days annually.

Table 4. Changing cropping patterns shifted the weed population from difficult-to-control perennials to easily controlled annual weeds. (Adapted from International Rice Research Institute, 1976a)

Season	Weeds (no./m^2)		
		Perennials	
	Annuals†	*Scirpus maritimus*	*Cyperus rotundus*
	Lowland rice followed by lowland rice		
Dry	725	406	0
Wet	340	325	0
	Corn + mung bean followed by lowland rice		
Dry	1008	36	32
Wet	364	167	0
	Corn followed by dry-seeded rainfed bunded rice		
Dry	2386	35	53
Wet	1463	20	33

† Grasses, broadleaved weeds, and sedges.

The high-yielding, short-season varieties of food crops, such as rice, wheat, and corn, recently developed by plant breeders, provide an obvious opportunity to increase cropping intensities in the tropics. Experiments conducted at various locations in South and Southeast Asia offer solid evidence that it is possible to develop management practices that will double and, in some cases, triple the number of crops that can be successfully grown annually. Examples of traditional and of intensive cropping systems are shown in Fig. 2.

During a three-week trip to the People's Republic of China in October 1976 and during a month's visit in September 1974, I was impressed in each province I visited with the extent to which the Chinese practice intensified cropping. A significant accomplishment in China of the last decade is the development and adoption of cropping systems which increase by at least one the number of crops grown annually.

In the Yangtze River Valley, where traditionally a summer crop of rice and a winter wheat or vegetable crop have been grown, the farmers are now producing two rice crops in summer, followed by a winter crop. The success of these intensified cropping systems serves as a model for scientists and researchers elsewhere.

The quick turnaround between the harvest of one crop and the planting of the next has contributed in large measure to the success of the crop intensification program in China. To illustrate the importance of quick turnaround, the Chinese describe a "brown-black-green" summer day. The crop is harvested (brown) in the morning; by early afternoon the soil has been prepared for the subsequent crop (black), and by evening a crop of rice has been transplanted (green). While such efficiency may not always be attain-

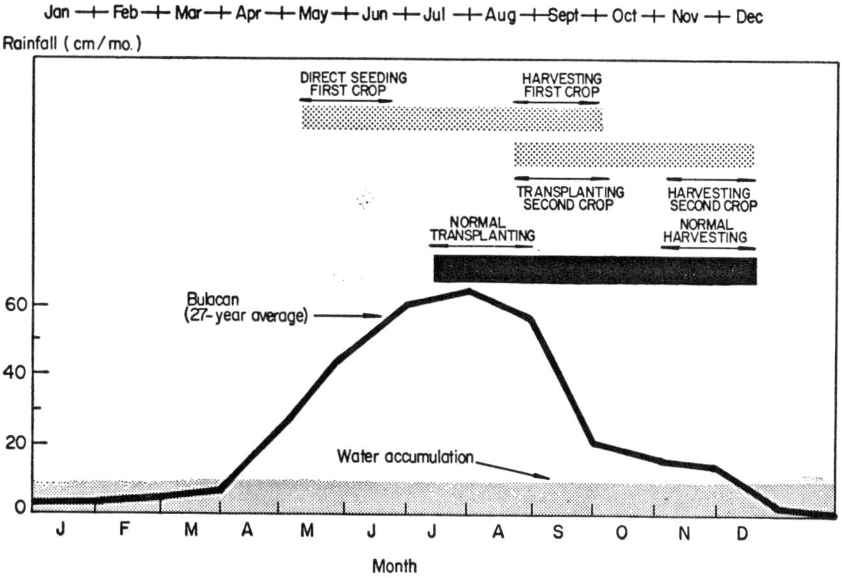

Fig. 2. A double cropping sequence (upper) and a traditional single cropping pattern (lower) in relation to rainfall distribution in Bulacan province, Philippines (International Rice Research Institute, 1975b).

able, it emphasizes the need to constantly keep a crop growing on the land to exploit the land's maximum crop production potential.

The Chinese experience suggests that the agronomist must give attention to management practices that quickly establish a second crop after the first is harvested. This work will require the efforts of a team of soil physicists, soil management specialists, agricultural engineers, and crop management specialists. Methods of tillage and of land preparation need to be developed to make it easy for the farmer to plant the crop quickly when the monsoon rains first come. Innovative methods of weed control must be developed for the humid tropics also, because primitive methods of planting may not furnish the crop seedlings the advantage they commonly have in temperate regions.

Agronomists must devote greater efforts towards developing methods for recycling wastes for crop production. In most countries in the tropics, the potential for waste recycling is far underutilized. There is relatively little use of the practice of composting as a mechanism of waste recycling. For example when I fly over areas of South and Southeast Asia immediately after rice harvesting and threshing, I am impressed by the extent to which massive quantities of rice straw are burned rather than being returned to the soil. In regions where inadequate levels of plant nutrients in the soil constrain crop yields, much of the nitrogen taken up and incorporated into the stem tissue of the previous crop is lost when the straw is burned. Except in East Asia (China, Japan, and Korea), residues are rarely recycled through compost. Attempts must be made to develop meaningful ways of recycling such agricultural waste.

Increased fertilizer prices will surely provide incentives for recycling techniques, but the leadership of agronomists is vitally needed. Soil and crop scientists should give attention to methods of composting. They must work in cooperation with pest management researchers to determine the relationship between recycling organic matter and the prevalence of and epidemiological ramifications to crop-damaging pests and their predators. Much can be learned by studying the successful waste recycling practices in Japan, Korea, and the People's Republic of China.

The increased worldwide interest in agricultural research should encourage agronomists and other agricultural scientists to continue and expand their efforts to improve world food production. A survey of expenditures for agricultural research and extension by Boyce and Evenson (1975) furnishes evidence of this increased interest (Tables 5 & 6), especially in Africa and Latin America. The expenditures remain low, however in Asia, where the food and population problems are most acute.

I would be negligent if I did not mention the role that agronomists can and should play in the education and training of scientists and educators in developing nations. Most agricultural research organizations in developing countries lack educated and trained manpower, especially qualified and experienced personnel who want to improve the food production problems of

Table 5. Public sector agricultural research and extension expenditures as percentage of the value of agricultural products in 1959 and 1974 (Boyce & Evenson, 1975)

	Research (%)		Extension (%)	
Subregion	1959	1974	1959	1974
Northern Europe	0.62	1.32	0.74	0.86
North American	0.85	1.27	0.43	0.55
Tropical South America	0.27	1.03	0.36	1.15
West Africa	0.68	1.12	0.80	2.46
South Asia	0.16	0.31	0.37	0.38
Southeast Asia	0.16	0.49	0.29	0.60

Table 6. Expenditures on research and extension as a percentage of the value of agricultural products by per capita income group in 1974 (Boyce & Evenson, 1975)

Country per capita income group (US$)	Expenditures (%)	
	Research	Extension
1750	2.55	0.60
1001-1750	2.34	0.31
401-1000	1.16	0.40
150- 400	1.01	1.59
150	0.67	1.82

their country. In some countries agronomists are classified with generalists and are not thought to require the same level of training as scientists in other scientific disciplines. Such priority rankings must be changed to prepare agronomists for their vital role in food production efforts.

The orientation and focus of the out-of-country education and training received by young scientists from developing countries should be reevaluated. The key word is relevance—relevance to the problems of the developing countries. The postgraduate education programs for those young scientists all too often have been essentially identical to those of students who expect to work in the more developed nations. Such programs are not designed to equip those scientists for the work they will undertake upon returning to their native land.

With the increased attention that the United States gives to world agriculture and with financial resources on the horizon to support this attention, American universities should carefully scrutinize the programs used for training students from developing countries. If possible, arrangements should be made to give a major share of the training in the geographic areas from which the student comes. For example, much of the thesis research work can be done there. Professors at American universities should design graduate programs that are relevant to the food production problems of developing countries. Title 12 will surely give support for the greater personal involvement of these professors in overseas work.

Agronomists have critical roles to play in increasing the world's capacity to produce more food. Their past performance is excellent, and their potential for future contributions is bounded only by their imagination, ingenuity, and diligence. But they must set their sights high. The impact of the semidwarf wheats and rices was due in large part to the dramatic yield increases that agronomists made possible. Agronomists should strive for similar quantum increases in the future because such increases are essential if world hunger is to be avoided.

LITERATURE CITED

Agricultural Board National Research Council. 1972. Genetic vulnerability of major crops. National Academy of Sciences, Washington, D. C.

Board on Agriculture and Renewable Resources, National Research Council. 1975. World food and nutrition study—enhancement of food production for the United States. National Academy of Sciences, Washington, D. C.

Boyce, J. K., and R. E. Evenson. 1975. National and international agricultural research and extension programs. Agricultural Development Council, Inc., New York, N. Y. 229 p.

Carlson, P. S., and J. C. Palacco. 1975. Plant cell cultures: Genetic aspects of crop improvement in food, politics, economies, nutrition, and research. Am. Assoc. Adv. Sci., Washington, D. C.

Chang, T. T., and B. S. Vergara. 1972. Ecological and genetic information on adaptability and yielding ability in tropical rice varieties. p. 431–453. *In* Rice breeding. International Rice Research Institute, Los Baños, Philippines.

Evenson, R. E. 1974. The Green Revolution in recent development experience. Am. J. Agric. Econ. 56(2):387-394.

Jennings, P. R. 1976. The amplification of agricultural production. Sci. Am. 235:180-194.

International Rice Research Institute. 1973. Annual report for 1972. Los Baños, Philippines. 246 p.

International Rice Research Institute. 1975a. IRRI research highlights for 1975. Los Baños, Philippines. 107 p.

International Rice Research Institute. 1975b. Two crops of rainfed rice. IRRI Reporter. Nov. 1975. Los Baños, Philippines.

International Rice Research Institute. 1976a. Annual report for 1975. Los Baños, Philippines. 479 p.

International Rice Research Institute. 1976b. Root-zone placement stretches scarce agricultural chemicals. IRRI Reporter. May 1976. Los Baños, Philippines.

Mighigan-Kettering. 1975. Crop productivity research imperatives. Michigan Agric. Exp. Stn., East Lansing, Mich., and Charles F. Kettering Foundation, Yellow Springs, Ohio. p. 260.

Poleman, T.T. 1975. World food: A perspective in food: politics, economies, nutrition, and research. Am. Assoc. Adv. Sci., Washington, D. C.